DICTIONARY OF PROPERTY AND CONSTRUCTION LAW

This dictionary provides a clear and concise explanation of the terms commonly used in land, property and construction law and management. The four key areas of coverage are; planning/construction law, land law, equity/trusts and finance/administration.

It serves as a useful reference for property and building professionals and for students of property and construction law on building, housing, construction management, surveying and law courses.

Jack Rostron MA DipLaw MRICS MRTPI is Senior Lecturer in Construction at Liverpool John Moores University. He is the author of several books and articles on property and construction law and sometime adviser to the World Health Organization. **Linda Wright** BA MBA is Programme Leader for Housing Studies at Liverpool John Moores University. She has worked extensively in housing practice and education. **Laura Tatham** BA MSc is Senior Lecturer in Law at Manchester Metropolitan University. **Robert Hardy-Pickering** DipLaw formerly with the State Law Office of the Government of Vanuatu.

DICTIONARY OF PROPERTY AND CONSTRUCTION LAW

Jack Rostron (Editor)
Robert Hardy-Pickering
Laura Tatham
Linda Wright

London and New York

First published 2002 by Spon Press
11 New Fetter Lane, London EC4P 4EE

Simultaneously published in the USA and Canada
by Routledge
29 West 35th Street, New York, NY 10001

Spon Press is an imprint of the Taylor & Francis Group

© 2002 Taylor & Francis

Typeset in Times by
Keystroke, Jacaranda Lodge, Wolverhampton
Printed and bound in Great Britain by
TJ International Ltd, Padstow, Cornwall

British Library Cataloguing in Publication Data
A catalogue record for this book is available from the British Library

Library of Congress Cataloging in Publication Data
A catalog record for this book has been requested

ISBN 0–419–26110–9
ISBN 0–419–26100–1 (hbk)

PREFACE

This work attempts to define and describe the words and terms in common usage by the property and construction community. In writing such a work, choices have to be made, especially in achieving a balance between comprehensiveness of individual entries and the number included. In striking this balance I have been conscious of the need to meet a range of potential usage.

I am grateful to my co-authors; Robert Hardy-Pickering (land law), Laura Tatham (equity and trusts) and Linda Wright (finance and administration), for their contributions. However, any shortcomings or criticism are my responsibility as editor.

I am grateful to Gower for allowing me to use material from some of my previous books.

I hope the book is of sufficient detail and coverage to be of practical use to practitioners and students of law and the landed professions.

Jack Rostron
February 2001, Liverpool

FOREWORD

In the ever-changing world that we now occupy, changing terminology, legalese and even sleaze, force the student and practised expert to reach for one or more books from the ever-growing body of reference work. The growth in litigation, especially concerning property and construction, has heightened the need for an awareness of the law. It is against this background that this concise dictionary was prepared, in the hope that it will provide a source of reference to the terms commonly used.

The book defines the words and phrases which are in common usage in construction, land transactions, housing, planning, equity/ trusts and related finance and administration. In terms of the definitions for each word or phrase, they have been carefully written in sufficient depth to provide the reader with an understanding of the principal issues and consequences.

Obviously, the book is modest in size. But it is essentially a reference handbook for students and practitioners. For this purpose the scale of the dictionary is one of its great virtues. Readers are not drowned in technical detail, as they can be when consulting traditional legal textbooks. On the contrary, simplicity, economy and clarity are the hallmarks of this dictionary; and it is the combination of these qualities which will ensure the reader will rapidly find what they want and that they will not be distracted by unnecessary detail. After all, if the reader wishes to pursue a point further, he or she can then turn to one of the standard textbooks for further assistance – an action which I feel would prove to be exceptional.

I congratulate the authors for its clarity and ease of use, whose compehensiveness belies its size and which will I believe in time become a standard reference for students of law, construction, and the landed professions, and it gives me great pleasure to commend it to readers.

Sir Joseph Dwyer FREng,
President of the Institution of Civil Engineers

A

abatement (1) Cancellation or reduction of a debt. For example, a lease, providing for the abatement of rent. (2) The abatement of an action is the bringing to an end or the suspending of an action. (3) The abatement of legacies refers to receipt by legatees of only a fraction or none of their legacy when assets are insufficient to satisfy all legacies. (4) The termination of a public or private nuisance.

abatement notice (1) A notice served on the owner or occupier of property regarding a private nuisance informing the person of the intention to abate the nuisance. (2) A statutory notice served under the Public Health Act 1936 by the local authority requiring a person to cease the nuisance.

abatement of nuisance An alternative to a legal action which an occupier of land may invoke by his own act abating any nuisance by which that land is injuriously affected, e.g. cutting off the branches of a tree. Notice is required if entry is necessary to adjoining land.

abatement of purchase-money Reduction of purchase price when a vendor misdescribes property and is unable to convey it as so described.

ab intestato From an intestate, meaning succession to the property of a person dying intestate, that is without a will.

abode The place where a person usually lives and sleeps.

abortive expenditure Investment, which has not achieved its purpose or has otherwise been wasted.

above par Stock exchange term applied to shares whose price exceeds their nominal value.

absolute Complete and unconditional, as in decree absolute.

absolute covenant A positive or restrictive covenant, sometimes called a qualified covenant.

absolute interest Full and complete ownership of property.

absolute performance Standard which it is impossible to improve upon. In manufacturing the theoretical quality standard zero defects.

absolute title Registered owner of land guaranteed by the state under the Land Registration Act 1925 which is the principal statute concerning the registration of an estate in land.

absorption costing A method of costing in which both fixed costs and variable costs are allocated to cost units, and total overheads are absorbed according to activity level.

absorption rate The rate calculated in an absorption costing system, in advance of an accounting period, for the purpose of charging the overheads to the production of that period.

abstract A summary or an abridgement of a legal document. Before the use of photocopying, public records were kept by abstracts of recorded documents.

abstract of title A summary of the public records relating to the title to a particular piece of land. A person may review an abstract of title to determine whether there are any title defects, which must be cleared before a buyer can hold clear title.

abut To border on, or adjoin another piece of land or property.

ACA Abbreviation for Associate of the Institute of Chartered Accountants in England and Wales.

ACAS Abbreviation for the Advisory Conciliation and Arbitration Service.

ACC Association of County Councils – representative and pressure group for rural authorities.

ACCA Abbreviation for Associate of the Chartered Association of Certified Accountants.

accelerated depreciation A term applied to a rate of depreciation of assets faster than the useful life basis normally applied.

acceleration (1) To bring forward a particular interest in possession or a reversion or a remainder before the time envisaged by the testator. The doctrine does not apply to contingent interests. (2) A future interest in land coming into possession. For example, the surrender of a lease by a tenant, thereby accelerating the landlord's reversion.

acceleration clause Provision in a mortgage document stating that if a payment is missed or any other provision is violated the whole debt becomes immediately due and payable.

acceptable use policy The rules governing acceptable behaviour on a particular portion of the internet.

access Approach or means thereof, e.g. where there is a right of access to a highway by the owner of adjoining land.

accessible dwellings Residential units specially designed for occupation by people with mobility impairment, and who may need to use a wheelchair on some occasions. (See mobility housing.)

accommodation agency A commercial organisation which provides details of housing accommodation available for renting which is controlled by the Accommodation Agencies Act 1953, as amended.

account A record kept in a ledger or on a computer file of all the financial transactions relating to one individual customer, supplier, asset or liability. (See current account, deposit account.)

accountability An obligation to give an account. Within limited companies it is assumed that the directors of the company are accountable to the shareholders. This responsibility is discharged in part by the provision of an annual report and accounts.

accountant An individual who has successfully completed the examinations of one of the recognised accountancy bodies and completed any required work experience. Work responsibilities will include collating, recording and communicating financial information.

accounting code Within accounting systems a numerical reference given to each account to facilitate the recording of accounting transactions.

accounting order A statutory instrument which provides the legal framework for housing association accounts.

accounting period A period for which a business or organisation prepares its accounts. When accounts are published the periods to which they refer must be stated and the beginning of one period should follow immediately upon the end of the preceding period.

Internally management accounts may be produced monthly or quarterly. Externally financial accounts are usually produced for a period of 12 months.

accounting principles The principles according to which accounts are prepared. The concepts adopted in the definition and calculation of individual items of income, expenditure, assets and liabilities. Such concepts have been agreed over time within the accounting profession and have more recently been the subject of various Statements of Standard Accounting Practice (SSAPs) to which members of accounting bodies are expected to conform.

accounting records Those records kept by an organisation in order that they may show and explain transactions and prepare proper accounts.

Accounting Standards Board (ASB) Body responsible for setting accounting standards in the UK. It was set up in 1990 to replace the Accounting Standards Committee (ASC) following the recommendations of the Dearing Report. The ASB is a subsidiary of the Financial Reporting Council.

accounts The profit and loss account and balance sheet of an organisation. (See annual accounts.)

accounts payable The amounts owed by a business to suppliers. These are classed as current liabilities on the balance sheet.

accounts receivable The amounts owing to a business from customers for an invoiced amount. These are classed as current assets on the balance sheet.

accrual In the accounts of an organisation an estimate of a liability which is not supported by an invoice or request for payment at the time the accounts are prepared. An accrual appears as a current liability on the balance sheet and will be charged under expenses in the profit and loss account.

accrual accounting System of accounting in which revenue is recognised when it is earned and expenses recognised as they are incurred. A basic accounting concept used in the preparation of the profit and loss account and the balance sheet of a business.

accruals concept One of the fundamental concepts contained in the Statement of Standard Accounting Practice. The concept that revenues and costs are matched one with the other and dealt with in the profit and loss account of the period to which they relate irrespective of the period of receipt or payment.

accrue   Literally to grow to, to increase, to be added as an increase, to come into existence.

accrued expenses   Costs relating to a period which have not yet been taken into account because they have not yet been invoiced by the supplier or been paid. These will include items which are generally invoiced in arrears.

accrued interest   Unpaid interest derived from a loan or an investment.

accumulated depreciation   A term used in published accounts for the total amount of the depreciation written off the cost price for valuation of a capital asset since it was brought into the balance sheet of an organisation.

accumulated profits   Within the appropriation of profits account the amount which can be carried forward to the following year's accounts after paying dividends, taxes and putting some to reserve.

accumulating shares   Shares issued in lieu of a dividend on ordinary share capital. Accumulating shares avoid income tax (but not capital gains tax) and are a way of replacing annual income with capital growth.

accumulation   When the interest of a fund comes into existence it will be invested as it accrues or arises. Restrictions have been imposed on accumulation partly by the Law of Property Act 1925 Sections 164–166 and partly by the rule against perpetuity. (See perpetuity.)

accumulation and maintenance settlement   A settlement in which there is no interest in possession but the beneficiaries will become entitled to an interest on attaining a specified age under 25 years.

accumulative rate   Investment rate interest which is assumed, or known, at which an annual sinking fund will grow.

acid test ratio   (See liquidity ratio.)

ACIS   Abbreviation for Associate of the Institute of Chartered Secretaries and Administrators.

acknowledgment   Formal declaration before a public official that one has signed a document. Prior to recording real estate legal documents, such as grant deeds and deeds of trust, a Notary Public acknowledges the person's signature on the document.

ACMA   Abbreviation for Associate of the Chartered Institute of Management Accountants.

acquiescence Agreement or consent, when expressed or implied, from conduct, for example silence or inactivity.

acquiring authority A government department, or local authority, using its statutory power of compulsory purchase.

acre A measure of land equal to 43,560 square feet.

ACS Arrears and credit statement.

act of God An incident resulting from a natural cause so devastating it is incapable of reasonable anticipation, e.g. earthquake, flood, landslide, etc.

action A civil legal proceeding started by a writ of summons.

action area An area designated by a local planning authority for comprehensive redevelopment, rehabilitation or development within a prescribed period.

action to quiet title A court action to establish ownership of real property. Although technically not an action to remove a cloud on title, the two actions are usually referred to as 'Quiet Title' actions.

active stocks Stocks and shares which have been actively traded on the London Stock Exchange.

active trust A trust which requires some active duties on the part of the trustee. (See bare trustee and trust.)

ADC Association of District Councils.

added value The value of improvement made on goods or services at any stage in their production. The difference in the price of a product or service taken in and the price at which the improved or finished product or service is sold either to the next internal or external consumer.

Additional Voluntary Contribution (ACV) Additional discretionary pension scheme contributions which employees can make in order to increase the benefits available from their pension fund on retirement.

ademption A specific legacy or specific devise fails by ademption if its subject matter has ceased to exist as part of the testator's property at her death. It does not apply to a general legacy nor a demonstrative legacy.

adjourn To put off the hearing of a case or matter to a later date.

adjustable rate mortgage (ARM) A mortgage where the interest rate is not fixed for the life of the loan. The interest rate adjusts periodically based on an index that changes with market conditions. The rate of interest is the sum of the index plus a margin (the margin remains fixed for the life of the loan). Most ARMs have periodic interest rate and payment caps, as well as a life cap.

administration Has several meanings. The affairs of a bankrupt can be said to be administered by his trustee in bankruptcy, the assets of a deceased's estate will be administered by the executor or the administrator.

administrator, feminine administratrix A person appointed by the court to manage the affairs and property of a deceased person.

ADP A term used by housing associations: Approved Development Programme.

ad valorem Literally according to the value. A duty calculated as a proportion of the value of the property to be taxed.

Advanced Corporation Tax (ACT) An advance payment of corporation tax paid when a company makes a qualifying distribution.

advancement (1) Power of advancement permits a trustee to find capital for the advancement or benefit of a beneficiary entitled to capital under a trust, see Trustee Act 1925 Section 32. (2) Presumption of advancement arises where there is a voluntary conveyance made to a wife or a child by the donor or to a person to whom he stands *in loco parentis* when the conveyance will be treated as an intended gift.

advancement clause A clause frequently inserted in wills or settlements permitting the trustee to release a fraction of a beneficiary's share for his advancement.

adverse possession Occupation of land without lawful title. It is possible to acquire title to land by proving 12 years' adverse possession under the Limitation Act 1980. Proof has to be shown not only that the occupier has (together with predecessors in title if necessary) been in possession for at least 12 years, but also that the possession has extinguished the title of the true owner.

advertisement Certain types of advertisements require permission under the Town and Country Planning Act 1990, as amended. The Secretary of State is empowered under this statute to make regulations controlling the display of advertisements, as far as it appears to be expedient in the interests of amenity or public safety. A wide range of advertisements is exempt from the need for formal consent, but permission is required for agents' boards above a certain size.

Advisory Conciliation and Arbitration Service (ACAS) A government body set up in 1975 by the Secretary of State for Employment to mediate in industrial disputes in both the public and private sectors. Its findings are not binding, but carry considerable weight in influencing both government attitudes and public opinion.

affidavit A written statement sworn on oath which may be used in certain cases as evidence.

affordability ratio Rent to income ratio.

affordable rent In the housing association context a term used in relation to assured tenancies as a means of keeping rent levels within the means of low income tenants.

after acquired property Property which is received later and can be dealt with in the form of a covenant to transfer.

Age Concern A charity established to look after the needs of the elderly.

agency A contract by which the agent undertakes to represent the principal in business transactions, using some degree of discretion.

agency board A board displayed outside premises advertising it for sale or to rent, usually containing the name of the estate agent. There are restrictions on the size and number of boards which can be displayed. The maximum permitted area for a board displayed on a residential property is 0.5 square metres for a single board and 0.6 square metres for two boards joined at an angle

agenda List of items to be discussed at a business meeting.

agent Person authorized to act on behalf of another in dealings with third parties.

AGM Annual general meeting.

agreement of sale An agreement between parties for the sale of real estate. Also may be known as a Purchase Agreement, Sales Agreement, or Land Contract.

a holding Land which is demised to a tenant.

alienable Capable of being transferred.

alienate To exercise the power of disposing of or transferring property.

alienation clause The specific provision in a mortgage document stating that the loan must be paid in full if ownership is transferred.

alimentary trust A protected trust.

allocation to a housing association The amount of money provided by the Housing Corporation for capital expenditure to be expended in the forthcoming year.

allotment A method of distributing previously unissued shares in a limited company in exchange for a contribution of capital. Application for such shares is often made after the issue of a prospectus on the floatation of a public company, or at privatisation of a state owned industry.

allotted shares Shares distributed to new shareholders by allotment.

allowances The estimated amounts a housing association will spend on maintenance and management. The estimate is used in the calculation of Housing Association Grant provided by the Housing Corporation.

alteration A major alteration, for example of a date which changes the sense of an instrument and generally invalidates it. An alteration in a deed is presumed to have been made before or at the time of the execution. An alteration in a will is presumed to have been made after the time of execution. See the Wills Act 1837 section 21.

alternate director A person able to act temporarily in the place of a named director of a company in his or her absence. The approval of other directors is a common requirement and the extent of the authority of the alternate director and his entitlement to remuneration will be determined by the relevant paragraph in the Articles of Association.

alternative dispute resolution (ADR) A phrase which describes various types of mediation and conciliation of a dispute without recourse to litigation.

altra duat comma in Literally in right of another. An example would be a trustee holds property in right of *cestui que trust* or an executor holds it in right of the deceased and his legatees. Also means grounds for some legal proposition, for example judicial decisions or opinions of authors.

AMA Association of Metropolitan Authorities.

amalgamation   The combination of two or more companies.

ambulatoria est voluntus defuncti us ad vicie supremum exitum   The will of a person who dies is revocable up to the last moment of life.

amenities   The qualities and state of being pleasant and agreeable. In appraising, those qualities that attach to property in the benefits derived from other than monetary. Satisfactions of possession and use arising from architectural excellence, scenic beauty and social environment.

amortisation   The writing-off of a wasting physical asset of its capital cost by means of a sinking fund.

amortisation rate   The rate of interest used for calculating amortisation.

amortisation term   The number of years applied to the useful life of an asset over which its value is written-off.

ancestor   One from whom a person is descended.

ancient lights   Right of access to light to a building enjoyed for 20 years without interruption, when the right becomes absolute.

Ancient Monuments and Archaeological Areas Act 1979   This Act requires the Secretary of State to prepare a schedule of monuments which appear to him/her to be of 'national importance'. (This could include almost any building structure or site of archaeological interest made or occupied by man at any time.) English Heritage and the Ancient Monuments Boards for Scotland and Wales would advise the Secretary of State. The fact that a monument is scheduled does not mean that it will be preserved at any cost; it ensures however that full consideration is given to the case for preservation if any proposal is made that will affect it.

ancillary use   A term used in town planning which describes the use of a property in a manner that is different from its main use.

animus cancellandi animus revocandi   The intention of cancelling or the intention of revocation especially in relation to a will.

Annual Abstract of Statistics   Annual publication of the Central Statistical Office giving UK industrial, legal and social statistics.

annual accounts   The financial statements of an organisation usually published on an annual basis. Copies of the annual accounts of incorporated bodies must be filed with the Registrar of Companies each year and must have attached to them a director's report

and auditor's report. Copies of these accounts must be sent to all
members and laid before members at an Annual General Meeting.
Annual accounts consist of balance sheet, profit and loss account,
cash flow statements if appropriate, and a statement of total recognised gains and losses. Companies falling into the legally defined
small and medium sized company categories may file abbreviated
accounts that may not have been audited.

annual general meeting (AGM) An annual meeting of the
shareholders of the company. An AGM must be held every year and
the meetings must not be more than 15 months apart. All members
must receive notice as prescribed in the Companies Act and that
notice must be accompanied by forms allowing a member to appoint
a proxy to attend and/or vote on his behalf. The usual business
transacted at an AGM is the presentation of the audited accounts,
the appointment of directors and auditors, the fixing of their
remuneration and recommendations for the payment of dividends.

annual percentage rate (APR) Where the rate of interest payable is
stated in terms of a rate per week, month or any period less than one
year then the annual percentage rate that is being charged is the
equivalent of the annual interest rate. This figure will be much higher
than the rate quoted for the short period. Most investment institutions
are required by law to specify the APR when interest intervals are
more frequent than annual.

annual return All companies are bound by the Companies Acts to
submit successive annual returns to the Registrar of Companies.
These documents must include the address of the Registered Office
of the company; the names, addresses, nationality of its directors.
Financial statements, directors' reports and auditors' reports must
be included. Unlimited companies are exempt from filing financial
statements. There are penalties for the late filing of accounts.

annual value The value placed on land for rating purposes. The gross
value is the rack rent, i.e. the rent per year on the open market, less
the landlord's costs incurred in paying for insurance, repairs, etc.
The net annual value, or rateable value, is the gross value less
statutory deductions.

annuity This is an annual payment of a sum of money, usually paid
via a pension scheme. Those created after 1925 are capable of being
registered as general equitable charges under the Land Charges Act
1972.

annul A declaration that court proceedings or their outcome no longer have any effect.

Anton Piller Order Takes its name from a case Anton Pillar KG v Manufacturer Processes Ltd 1976 1 All ER 779 CA and it requires a defendant to allow the plaintiff to enter his premises and remove documents.

appeal A referral to a superior authority or court, or a judicial or administrative review of the decision, by an inferior body. An example would be an appeal to the Secretary of State against refusal of planning permission.

appearance Acknowledgement, by the defendant in a civil action, of the writ of summons.

appellant The party who brings an appeal to a higher court.

appendant A subordinate interest or right attaching to a larger interest in land which, by virtue of law, will automatically pass with the conveyance of the greater interest.

appointed day The day on which an Act of Parliament comes into operation.

appointment, power of This is the power given to a person usually by a trust or settlement enabling them to dispose of property not their own. (See power.)

apportionment This is the division of a legal right into its proportionate parts according to the interests of the parties involved.

appraisal report Estimate of value. An appraisal evaluates the property at a given time based on facts regarding the location, improvements, neighbourhood and comparable sales.

appraised value An expert opinion of the value of a property at a given time, based on facts regarding the location, improvements, etc., of the property and surroundings.

appraiser A person who determines the value of property. Normally called a valuer in the United Kingdom.

appreciation An increase in the value of an asset, particularly a fixed asset such as land or buildings. This appreciation often occurs as a result of inflation. The directors of a company have an obligation to adjust the value of land, buildings and other assets in balance sheets to take account of appreciation.

appropriation (1) The allocation of a sum of money for expenditure. (2) Any part of the real or personal estate of a deceased person can be appropriated in or towards satisfaction of a legacy or a share of residue.

appropriation account The allocation of the net profit of an organisation in its accounts. Payments such as wages, salaries, heat and light and interest payments will be treated as expenses and deducted before arriving at net profit. Other payments such as dividends to shareholders, transfers to reserves and amounts for taxation will be deemed to be appropriation of profit once that profit has been ascertained.

Approved Development Programme (ADP) The Housing Corporation's cash limit for capital expenditure on different types of project for each financial year, approved by the Secretary of State for the Environment. After this the Housing Corporation allocates its funds to individual regions and housing associations. ADPs are also granted by the Secretary of State to some local authorities which fund programmes of housing association activity.

approved inspector Under the Building Regulations a person who is approved by the Secretary of State or a designating body to supervise building work under powers conferred by the Housing and Building Control Act 1984, as amended. The statute establishes standards concerning the construction of buildings. All building work has to meet these standards and be approved.

Approved Landlord Scheme A local authority maintained registration scheme for private landlords who meet predetermined criteria.

AR19 A proforma which housing associations registered as Industrial and Provident Societies are required to complete and return each year to the Registrar of Friendly Societies.

arbitrage The deliberate switching of funds between markets in order to maximise gains on short-term investments.

arbitration The determination of a dispute by a third party. A procedure favoured in property disputes avoiding need for normal litigation process. Arbitrations are of three types: statutory, commercial and county court. The decision of the arbitrator is binding except in exceptional circumstances.

architect A person who designs buildings and is qualified by virtue of the Architects Registration Act of 1938, as amended. (See design team.) It is an offence under the Act for a person to call themselves an architect unless their name appears on the register maintained by the Architects Registration Council of the United Kingdom, which is the regulatory body established under the Act.

architect's certificate The certificate issued at various stages of construction by an architect, normally in accordance with the release of funds to a contractor on completion of various stages of work. (See quantity surveyor.)

architect's instruction A written instruction specifying alterations or additions to the building contract. (See variation order.)

ARCUK Architects Registration Council of the United Kingdom. It is unlawful for a person to call themselves an architect unless their name appears on the register.

area of archaeological importance An area designated under the Ancient Monuments and Archaeological Areas Act 1979, as amended, being one of archaeological importance so designated by the Secretary of State or local authorities. It is an offence in such areas to carry out, or cause or allow to be carried out, operations which disturb the ground or involve flooding or tipping without serving an operations notice on the local authority or unless specific exemption is granted.

ARLA Association of Residential Lettings Agents.

arrears Unpaid sums of money, normally rent that has not been paid on the due date. The terms for payment are normally stated as set periods in the lease agreement. Interest is said to be paid in arrears since it is paid to the date of payment rather than in advance.

articles of association Document setting out the voting rights of shareholders, conduct of shareholders and directors' meetings and the internal regulations governing the running of a company which is registered when the company is formed. The articles are subject to the memorandum of association and must not contain anything illegal or *ultra vires*. The articles may be altered by the company by special resolution at a general meeting.

ASB Accounting Standards Board.

'as is, where is' A clause that is sometimes used in the transfer of property. It means that the present property is being transferred with no guarantee or warranty provided by the seller.

asking price The price stated by the vendor of a property which is placed on the market for sale. The eventual price may be less than the asking price.

assembly of land The assembling of individual parcels of land to form a larger unit normally for the purpose of development of the greater unit created.

assent Consent. A document which acknowledges the right of a legatee or a devisee to property under a will. An assent to the vesting of a legal estate must be in writing, signed by the personal representative, and must name the person in whose favour it is given.

assessed valuation Value placed on real estate by governmental assessors as a basis for levying property taxes; not identical with appraised or market value.

assessment The determination of a person's liability for various taxes, normally undertaken by an Inspector of Taxes.

assessment base The total assessed value of all property in a given assessment district.

asset valuation The determination by expert opinion of the financial worth of the property which is often incorporated into company accounts.

assets Property or rights having a monetary value. Property which is available for paying debts. In terms of Capital Gains Tax, the form of property including options, debts, and other forms of property created by the person disposing of it.

assign To transfer property to another by assignment.

assignee The person who takes the rights or title of another by assignment.

assignment The transmission by agreement of a right or interest or a contract to another person. For example, the situation where a tenant sells his lease. The tenant selling the lease is called the assignor, and the person who is buying the lease is called the assignee. All that the tenant can sell is what remains of his lease – the unexpired residue of the term.

assignor A person who transfers property rights or powers to another by assignment.

association investment profile The Housing Corporation summary record of a housing association's overall performance. It forms part of the investment file used by Housing Corporation staff to establish whether a particular housing association is eligible for further capital funding.

Association of Metropolitan Authorities (AMA) Pressure group for metropolitan authorities.

assured shorthold tenancy A tenancy granted under the Housing Act 1988 for a minimum of six months. The landlord cannot determine the tenancy during the initial six months.

assured tenancy A tenancy under the Housing Act 1980, as amended, where the tenant has security of tenure.

asylum seeker Person fleeing from persecution or oppression abroad who has applied for refugee status within the United Kingdom.

at best Instruction to a broker to buy or sell shares, stocks, commodities, currencies, etc. at the best possible price.

at call Money which has been lent on a short-term basis and must be repaid on demand.

attached homes A home that has one or more common walls adjoining another home. Flats, semi-detached and terraced houses are attached homes.

attendance allowance Allowance under the Social Security Act 1975, as amended, payable to disabled people who require constant attendance.

attestation Authentication of a document by the signatures of witnesses.

attestation of will The Wills Act 1837 requires the signature of two witnesses who are not a party to the will.

attornment Transfer of property following a sale.

auction A system where an auctioneer offers property for sale, selling to the highest bidder. A contract is made when the hammer falls in response to a final highest bid.

auctioneer A person who conducts an auction.

audit Independent inspection of financial statements of an organisation to ascertain whether or not the accounts are properly kept and show a true and fair view of the state of affairs of the organisation at the date stated. External audits (those performed by an auditor external to the organisation) are required under statute for limited companies by the Companies Act and for various other undertakings such as housing associations and building societies by other Acts of Parliament. Internal audits are performed by auditors within an organisation, usually an independent department. Internal auditors will examine both financial and non-financial areas with the aim of ensuring that internal controls are working effectively.

Audit Commission Government body with responsibility for ensuring value for money within local authorities for the National Health Service and the police. Since the Housing Act 1996 the Audit Commission has responsibility to work with the Housing Corporation examining the performance of registered social landlords.

auditor A person or firm appointed to carry out an audit of an organisation.

auditor's report A report made by an auditor appointed to examine and report on the accounts of an organisation. The report may take different forms, but auditors of a limited company are required to form an opinion as to whether the annual accounts of the company give a true and fair view of its profit or loss for the period and its state of affairs at the end of the period. They are also required to certify that the accounts are prepared in accordance with the requirements of the Companies Act (1985). The report is technically made to the members of the company and must be filed, together with the accounts, with the Registrar of Companies under the Companies Act (1985).

authorised capital The maximum amount of share capital which may be issued by a company. This will be detailed in the company's Memorandum of Association.

authority Power given by one person to another allowing the latter to do some act. See the Trustee Act 1925 section 23.

average cost The average cost of producing goods often referred to as the unit cost. This is calculated by dividing the total costs both fixed and variable by the total units of output.

avoidance Setting aside, making null or void.

award An award normally made by an arbitrator after the consideration of the dispute between two parties, for example, rent review.

B

back to back credit   Method used to conceal the identity of the seller from the buyer in a credit arrangement.

back to back loan   Loan from one company to another in a different country using a bank or other financial institution to provide the loan, but not the funding which comes from a third party. A simultaneous transaction of property from one party to another secured by way of a loan.

backland   A parcel of land which has no frontage to a highway.

BACS   Abbreviation for bankers automated clearing system. BACS provides an automated clearing house service in the UK.

bad debt   A debt which it is known, assumed or expected will not be settled. The full amount should be written off to the profit and loss account of the period or to a provision for bad and doubtful debts as soon as it is foreseen on the grounds of prudence. Bad debts which are subsequently recovered should be written back to the profit and loss account or to a provision for bad and doubtful debts.

bailiff   An officer entrusted with the execution of writs and processes.

bailment   Delivery of goods into the possession of a person who is not their owner.

balance sheet   Statement of the financial position of an entity at a given date disclosing the value of the assets, liabilities and accumulated funds such as shareholders' contributions and reserves which is prepared to give a true and fair view of the state of the entity at that date. The Companies Acts requires that the balance sheet of a company must give a true and fair view of its state of affairs at the end of its financial year and must comply with statute as to its form and content.

balloon loan   Mortgage in which the remaining principal balance becomes fully due and payable at a predetermined time. Most of the time, balloon loans have level payments until the note becomes due and payable.

bank advance   An amount lent by a bank to its customer as either a bank loan or an overdraft. The advance will usually be made for a specified time at a specified rate of interest. In most cases banks will require some form of security.

bank guarantee An undertaking given by a bank to settle a debt should the debtor fail to do so.

bank rate The rate at which the Bank of England was prepared to lend to other banks within the banking system until 1972. From 1972 to 1981 the term 'bank rate' was discontinued and replaced by 'minimum lending rate' (MLR). It is now more usually known as the base rate.

banker One who receives money from customers to place on deposit and pays out following instructions from his customer. Usually an incorporated body which is regulated in the conduct of its business by the Banking Act 1987.

bankrupt A debtor who has been adjudged bankrupt, indicating a failure to meet his or her liabilities. Their assets are vested in a trustee for bankruptcy following a declaration of bankruptcy in court.

bankruptcy The condition of a person judged to be insolvent, and the appointing of a trustee to administer their affairs.

bankruptcy, trustee in A person appointed by the bankrupt's creditors or by the Secretary of State or by the court. Once appointed, the bankrupt's estate vests in the trustee immediately.

bar The profession of barristers from which practically all judges are recruited.

bare trust A trust which does not require any active duties on the part of the trustee, but merely to hold the property in trust and to hand over the property to the person entitled to it at that person's request.

bare trustee A trustee who has no active duty other than to convey the estate to his beneficiary or according to the beneficiary's directions.

barrier free Design principles which attempt to achieve a high degree of accessibility for physically disabled people.

base rate The minimum rate of interest demanded by a bank on money loaned. This will be lower than the actual rate which will be fixed according to market pressures and the element of risk involved. This term is also applied to the rate at which the Bank of England lends to discount houses.

base rent The lowest rent stated in and payable under a lease.

basis Original cost of property plus value of any improvements put on by the seller minus the depreciation taken by the seller.

basis point One hundredth of 1%. A term often used in finance when price involves fine margins.

batch processing Method of processing data in which data/programs are collected into groups or batches for processing. Batch processing is frequently used for tasks such as payroll preparation.

BCA Basic Credit Approval. Term introduced by the Local Government and Housing Act 1989 to define the borrowing which a local authority may make in order to undertake capital projects.

bear Stock exchange term referring to a speculator who sells shares to be delivered at a future date anticipating that the price will fall and that the promised shares can be acquired at a price lower than the selling price when the time for completion of the contract arrives. The difference between the purchase price and the original sale price represents the successful bear's profit. A bear market is one is which the dealer is more likely to sell than buy.

bearer securities Stocks, shares and debentures for which no register of ownership is kept by the company. These are therefore transferable by hand, being made out to the bearer and not to a named person. Dividends are not received automatically from the company, but must be claimed by removing and returning 'coupons' attached to the certificates.

before and after valuation The process under which property is valued in an existing state and thereafter when the property has been changed in some material way. Often used in association with town planning compensation, following a Revocation Order which is issued by the local planning authority rescinding a previously granted planning permission.

below the line A term denoting entries below the horizontal line on a company's profit and loss account separating entries that establish profits from entries which show how the profit is distributed (or where the funds to finance the loss have come from).

benchmarking Identification of best practice in relation to products and processes from within and outside an industry with the aim of developing a guide and reference point for improving the practice of an organisation. Benchmarking may take place within an organisation, within an industry or sector or in relation to organisations in totally different fields.

beneficial interest The equitable interest of a beneficiary. The right of enjoyment or equitable interest as opposed to the nominal ownership or legal interest held by the trustee.

beneficiaries, remedies of. These are rights of a beneficiary arising where a trustee breaches the terms of a trust.

beneficiary (1) Somebody entitled for his or her own benefit. Known also as the *cestui que trust*. (2) A person who receives a gift under a will. (3) The lender named on the mortgage note.

beneficiary principle This principle arises in the creation of a private express trust. No private express trusts will be valid unless there is a human ascertainable beneficiary in whose favour the court can order performance, see Re Astor's Settlement Trusts (1952) 1 All ER 1067 and Morice v Bishop of Durham (1804) 7 RR 232.

Benefits Agency Government agency with responsibility for the allocation of social security benefits.

benefits in kind Benefits other than money received by employees by reason of their employment. Benefits in kind may be taxable particularly those received by directors and higher paid employees.

Benjamin Order An order made by the court authorising distribution of a person's estate based on the assumption that a beneficiary is dead because he cannot be found by the person's representatives and they cannot finish winding up the estate.

bequeath To leave or dispose of personal property by will.

bequest A disposition by will of personal property; a legacy.

BES Abbreviation for Business Expansion Scheme.

better equity Where somebody in the eyes of a court of equity has a better claim over property of another person. That person is said to have a better equity.

bill of sale Document transferring ownership of goods to another. The goods may be transferred conditionally as security for a debt. A conditional bill of sale is therefore a mortgage of goods and the mortgagor has the right to redeem the goods on repayment of the debt. An absolute bill of sale transfers ownership of the goods absolutely.

binder Preliminary agreement of sale, usually accompanied by earnest money (term also used with property insurance).

black and minority ethnic housing association strategy A strategy devised by the Housing Corporation in 1986 which aims to promote black and minority ethnic housing associations.

blanket mortgage A mortgage covering more than one property of the mortgagee.

blanket policy A policy which covers many different types of risk but has only one total sum assured. Blanket policy may cover a fleet of vehicles or a group of buildings.

block A properties Housing developments built before the introduction of the Housing Act 1988 Housing Association Grant, which are not eligible for major repair housing association grant funding and have to rely on sinking funds set up from rent surplus fund transfers.

block B properties Housing schemes developed prior to the Housing Act 1988 which are still eligible for major repair housing association grant funding.

block plan A simple plan showing the delineation of buildings, and other physical features, on a particular site. (See location plan.)

block width The dimensions across a building measured wall-to-wall.

blue chip Term applied to organisations which are seen as low risk investments producing a high return.

boarding house A private dwelling used for the purposes of short-term accommodation, which also provides meals, cleaning, etc.

bona fide A genuine act in good faith.

bond A term widely used to describe fixed interest securities issued by government, local authorities or companies. Bonds fall into various categories with fixed or variable rates of interest, redeemable or irredeemable, short or long term, secured or unsecured and marketable or unmarketable.

bonus dividend Dividend issued to a shareholder in addition to those expected.

bonus shares Shares issued to existing shareholders of a company. The number of shares received depends on the level of the shareholding prior to the bonus issue.

book of account The books in which a business records its transactions.

book value Accounting term describing the recorded value of an asset in the books of the business.

boom The point in the trade cycle at which the economy is working at full capacity. Demand, prices and wages rise while unemployment falls.

borrowing facility Arrangement with a financial institution which allows an organisation access to borrow money.

boundary The delineation by way of a line showing the separation of property in two separate ownerships.

breach Non-fulfilment of some contractual obligation.

breach of contract The non-fulfilment of the terms of a contract. The remedy is normally damages, specific performance, or recision.

breach of trust This may arise from failure to carry out the trustee's general duties or some abuse of his or her other powers. A trustee is normally liable for any loss caused directly or indirectly to the trust property and to the beneficiary's interest as a result of the breach, see the Trustee Act 1925 Section 61 and the Limitations Act 1980 Section 21.

breach of warranty Where someone fails to comply with a contractual agreement.

break clause The clause in a lease which gives the tenants and/or the landlord the right, in certain circumstances, to terminate the lease before its normal termination date.

break even analysis Management accounting technique in which costs are analysed according to cost behaviour characteristics into fixed costs and variable costs and compared to sales revenue in order to determine the level of sales volume, sales value or production at which the business makes neither a profit nor a loss.

break even point The level of production, sales volume or sales revenue at which an organisation makes neither a profit nor a loss.

break point The date on which a lease terminates when a break clause is enforced.

bridge financing A form of an interim loan generally made between a short-term loan and a long-term loan when the borrower needs additional time before obtaining permanent financing.

bridging interest (1) Interest charged to cover a bridging loan between sale and purchase of properties. (2) Interest chargeable on loans to cover expenditure by a housing association during the development of a housing scheme. The interest is normally funded from on-costs.

bridging loan A temporary loan taken on a short-term basis and used to bridge the gap between the purchase of one asset and the sale of another. It is particularly common in house or property purchase.

bridleway Highway over which the public has a right of passage on foot or horseback.

British Institute of Management An organisation set up by the government in 1974 to promote good practice in management and to disseminate information.

British Standards Institution An organisation established by Royal Charter in 1929 with a view to regularising standards of quality in materials, products and practices. Codes of practice established by the Institution ensure good standards of practice with regard to the manufacture and installation of materials, and design and workmanship. The majority of materials used for the construction of buildings are subject to British Standards with regard to testing and installation.

broker Agent who brings together buyer and seller for commission. There are specialist brokers, such as insurance brokers and stock brokers who deal in services rather than goods. The broker's remuneration consists of a brokerage usually calculated as a percentage of the contract sum.

brothel A house used for the purposes of fornication by both sexes. It is an offence to keep a brothel.

brown field site Site which has been previously used. Such sites are often covered by derelict buildings or suffer from land contamination.

BSA Building Societies Association.

budget A financial plan prepared for a future period, usually one year. Based upon the organisation's objectives the budget will contain the plans and policies to be pursued during the period. Constituent budgets may relate to costs, revenues, working capital movements, capital expenditure and cash flows.

budgetary control A system of financial control exercised within an organisation using budgets for income and expenditure for each function of the organisation in advance of an accounting period.

Budgets are periodically compared with actual performance to establish variances.

buffer stock The stock of a commodity which is bought, recorded and sold in order to stabilise market prices.

buffer zone A landscaped area separating developed areas. It is usually created to increase amenity and provide informal recreational space.

builder A person or organisation who is responsible for constructing a building in accordance with the information provided by the architect. (See contractor.)

builder's quantity surveyor (See contractor's quantity surveyor.)

building agreement A contract between the owner of land and a developer, normally in the form of a licence where the developer agrees to construct buildings and on the due date is entitled to a lease of the land and buildings.

building bye-laws (See by-laws.)

building code Government regulations specifying minimum construction standards.

building contract A contractual agreement between the owner or occupier of land and a building contractor, indicating the terms and conditions under which the building is to be constructed.

building contractor Someone who enters into a contractual agreement with an employer for the purpose of carrying out building or engineering works specified in the contract over the prescribed period.

building control The process by which local authorities, or other approved agencies, approve and supervise the construction of works in accordance with the Building Regulations.

building depth Maximum external dimension from the front to the rear walls of the building.

building frontage (See frontage.)

building lease A lease usually for 99 years for a rent known as ground rent where the lessee agrees to erect certain buildings which become the possession of the lessor on expiration of the lease.

building line The delineation of a fixed distance from the centre of a highway in front of which building is not permitted. The building line may be established by a filed plan of subdivision, by restrictive covenants in deeds or leased, by building codes, or by zoning ordinances. It is normally intended as a means to enhance the built environment and increase road safety.

building regulations A statutory instrument which lays down the approved methods of construction. They are administered by local authorities and other specified organisations. The Building Regulations 1991, as amended, cover the following matters: Structure, Fire safety, Site preparation, Toxic substances, Sound, Ventilation, Hygiene, Drainage and waste disposal, Stairs, Access for the disabled, Glazing, Heat and Energy.

building services engineer A specialist involved in the design and commissioning of services to a building such as heating, ventilation, electricity, water supply, drainage.

Building Societies Act 1986 This is the regulating legislation for building societies. This Act gave building societies new, wider powers directly to provide housing for rent and sale (as well as extending societies' lending powers).

building society An organisation established for the purposes of raising funds from its members to allow advances to be made to other members to purchase and build residential properties. They are non-profit organisations controlled by the Building Societies Commission and registered with the Registrar of Friendly Societies. Building Societies are a major lender to housing associations.

building survey A report of the condition of an existing building carried out and prepared by a building surveyor.

building surveyor (1) Someone who undertakes surveys of buildings. (2) A qualified person who is mainly responsible for dealing with existing building stock in terms of maintenance, alterations, upgrading, conversion and adaptation. (See surveyor.)

buildings of special architectural or historic interest Buildings which are listed of special architectural and/or historic interest under the Planning (Listed Buildings and Conservation Areas) Act 1990, which requires them to be subject to a special form of planning control intended to prevent the unrestricted demolition, alteration or extension without express consent from the local planning authority.

bulk buying The practice of buying products in sufficiently large quantities to take advantage of bulk discounts or other quantity discounts.

bull A stock exchange dealer who expects prices to rise. They contract to buy at a fixed price in advance at a future date in anticipation that when the contract matures the price on the market will have risen.

bungalow A single-storey residential dwelling.

business cycle Cyclical nature of economic activity in which periods of boom with increased investment output and employment are regularly followed by periods of recession.

business expansion scheme An arrangement which allows persons to secure tax relief after investing in a company involved in a particular type of business activity. Shares must be held for a minimum of five years to qualify for exemption from Capital Gains Tax.

business plan Document which sets out an organisation's plans for its future operations and development.

buyer's broker Agent who takes the buyer as a client, is obligated to represent their interest above all others, and owes specific fiduciary duties to the buyer.

buyer's market (1) A situation in which the buyer has an advantage in purchasing property. (2) Where demand exceeds supply.

by-laws or **bye-laws** Rules made by authorities for the administration, management and regulation of a particular area, undertaking, etc. and binding on all persons coming within their scope. By-laws are usually the means by which local authorities exercise some of their regulatory functions. By-laws usually require the approval of the Secretary of State responsible.

C

CAB Citizens Advice Bureau.

calendar month One of the periods of 28, 30 or 31 days in which a calendar year is divided. For leap years, a period of 29 days occurs in February.

calendar year A period of 365 consecutive days, or in a leap year, 366 consecutive days including the 29th February.

call loan A repayable loan, redeemable in full on demand.

called up capital Proportion of the issued share capital of a company that had actually been called up.

calls In company law a term used for the demands from a company to its shareholders to pay a certain sum in respect of their shares.

capacity A person's ability to enter into a legally binding agreement including contract and trusts.

capital Accumulated wealth. A portion of wealth which is set aside for the production of additional wealth. Specifically, the funds belonging to the partners or shareholders of a business invested with the expressed intention of them remaining permanently in the business.

capital adequacy Test of whether an organisation has the minimum value of reserves to run its business.

capital allowances Allowances set against income tax or corporation tax available to organisations for the depreciation of capital assets.

capital appreciation Increase in the value of a capital asset over a stated period, which is normally expressed as a percentage of the original value.

capital asset (fixed asset) Assets of material value that are long life and are held to be used in the business rather than for re-sale or conversion into cash. Examples of capital assets would include land, buildings, plant and machinery.

capital budget The sum allocated by an organisation for future capital expenditure.

capital expenditure Sums of money spent on improving and acquiring capital assets as distinct from revenue expenditure on such matters as maintenance.

capital gain In terms of Capital Gains Tax, the increase in value of capital assets. Under legislation Capital Gains Tax is payable when certain items are disposed of and at certain occasions.

capital improvement Capital works undertaken on an asset with a view to enhancing its value. It does not include repairs or maintenance.

capital market Market in which capital is raised by industry, government and local authorities. Funds are normally raised through the issue of shares, bonds, debentures, loan stock, etc.

capital money   Funds paid to trustees under the Settled Land Act 1925. Under the Act settled land is: (1) Limited in trust to any person by way of succession. (2) Limited in trust for any person in possession. (3) Limited in trust for a contingent estate. (4) Limited to or in trust for a married woman. (5) Charged with any rent charge for the life of any person. The instrument by which land is settled is called the settlement.

capital receipts   In respect of local authority housing money received from the sale of capital assets, mainly council housing and land. Currently central government controls how such funds may be used.

capital value   The capital value of an asset, as opposed to an annual or periodic value, such as rent.

capitalised interest   Bridging interest accumulated on loans released during the development period. (See bridging interest.)

capite, tenure in   Holding land directly from the Crown.

capricious   A concept developed by Templeman J in Re Manisty's Settlement (1974) Ch 17 in which he upheld a power of very considerable breadth. His example of a capricious power was one in favour of the residents of Greater London which was, he suggested, invalid because the terms of the power negative any sensible intention on the part of the settler and any sensible consideration by the trustees of the exercise of the power. The concept applies to both discretionary trusts and powers.

caps   Used on adjustable rate mortgages (ARMs) to limit the interest rate and/or the payment. Most ARMs have a periodic cap that is around 2% per year and a life cap of around 5%–6% over the life of the loan. 'Payment only' caps sometimes create negative amortisation where the principal balance of the loan increases rather than decreases over time.

caravan   A mobile home which is capable of being moved from site to site. Normally used for short periods of time as a holiday residence.

Caravan Sites and Control of Development Act 1969   This Act gave local authorities new powers to control caravan sites, including a requirement that all caravan sites had to be licensed before they could start operating (thus partly closing loopholes in the planning and public health legislation). These controls over caravan sites operate in addition to the normal planning system; so both planning permission and a licence have to be obtained. Most of the Act dealt with control, but local authorities were given wide powers to provide caravan sites.

care and repair A colloquial term describing agency services to assist elderly home-owners with housing maintenance, adaptations, etc. (See staying put.)

care contract An agreement making provision for care arrangements between two parties, e.g. social services and registered social landlords. Such agreements are enforceable by law.

care in the community A central government policy for releasing certain classes of inmate or patient from hospitals and institutions and arranging for care in the community.

carriageway A highway, along which there is a right to drive vehicles and possibly to lead or ride animals and to pass on foot, unless exempt by statute.

case A court action, often referring to the particulars of a specific action.

case stated A statement of facts prepared by a court or arbitrator for the decision of a higher court on a point of law.

cash back A sum of money paid by a property owner to a potential tenant in times of over-supply.

cash book In book keeping the book of prime entry where all receipts and payments are recorded. Receipts are shown as debits and payments as credits.

cash cow A term deriving from the Boston matrix to describe a product with a high market share within an established market which produces a steady flow of cash.

cash flow The movement of money both in and out of an organisation within a specified period.

cash flow forecast Forecast of expected payments and receipts within a given period. Although an organisation may have a healthy reserves and assets position it may experience severe problems if it does not have sufficient to cover cash commitments.

cash limit Limits on actual spending. In the case of social housing the Department of Environment, Transport and Regions (DETR) will set a cash limit for the Housing Corporation each year.

casting vote A deciding vote. In common practice the vote given to the chair of a meeting to be used if the votes cast for and against a particular resolution are the same. The arrangements concerning the right to casting vote are usually contained in the Articles of Association.

catch up repairs Repairs required to bring a property into good condition.

category 1 housing Defined in circular MHLG 82/69 as . . . 'self contained dwellings to accommodate one or two older people of a more active kind.' This circular is no longer mandatory.

category 2 housing Defined in circular MHLG 82/69 as . . . 'accommodation in grouped flatlets to meet the needs of less active older people.' The circular is no longer mandatory.

caution The Land Registration Act 1925 allows any person interested in land to lodge a caution with the Registrar, requiring notification of any dealings in the land. The Act introduced a system whereby all land in England and Wales is subject to compulsory registration on first conveyance or sale of the freehold or grant of a lease for more than 21 years.

caveat A warning.

caveat actor Let the doer beware.

caveat emptor Let the buyer beware; a maxim indicating that any risk is upon the buyer and not the seller.

caveat venditor Let the vendor beware.

CBI Abbreviation for Confederation of British Industry.

CECODHAS European Liaison Committee for Co-operative and Social Housing. It represents national member organisations in 16 European countries which are providers of social housing.

ceiling value The maximum compensation payable for compulsory acquisitions, normally associated with site value for unfit houses.

Central Statistical Office UK government department which collects and collates information concerning the national economy and reproduces this information in statistical form.

CEO Abbreviation for chief executive officer.

certainties, the three The necessary conditions for the creation of a valid private trust stated by Lord Eldon in Wright v Atkyns

(1823) Turn & R 143 and by Lord Langdale in Knight v Knight (1840) 3 Beav 148.

Certificate of Incorporation Certificate issued to the shareholders of a company by the Registrar of Companies which brings a company into existence. It is issued following the submission of Memorandum and Articles of Association together with other documentation. A company has no legal existence until the certificate has been issued.

certificate of making good defects The certificate issued by an architect or surveyor under a building contract at the end of the defects liability period, indicating that the contractor has satisfactorily made good any defects. It has the effect of releasing the balance of retention money to the contractor.

certificate of origin Documentation stating the country of origin of any exported goods. The certificates are usually issued by the Chamber of Commerce in the country of origin and determine whether or not an import duty has to be paid and if it has, at what tariff.

certificate of practical completion A certificate issued by an architect or surveyor indicating that works have been substantially completed and the building is ready for occupation. The certificate allows the release of an agreed percentage of retention monies.

certificate of value A certificate signed and issued by the purchaser of property attesting the price paid, often for the purposes of stamp duty.

certificate, land A document under the seal of the Land Registry which contains the particulars of a piece of land.

Certified Accountant Member of the Chartered Association of Certified Accountants, one of the major bodies of accountants whose members are recognised as properly qualified to audit the accounts of limited companies.

certified copy A true copy, attested to be true by the officer holding the original.

cesser The premature cessation of a right or interest.

cessio bonorum The surrender by a debtor of his property to his creditors.

cestui que trust The beneficiary. One having an equitable interest in property with the legal title being vested to the trustee. That is, the person for whose benefit a trust is created.

cestui que use   One to whose use the property was conveyed.

cestui que vie   The person for whose life any land is granted. Thus, where 'X' is a tenant for the life of 'Y', 'Y' is a cestui que vie.

chain of title   The chronological order of conveyance of a parcel of land from the original owner to the present owner.

chain transactions   The situation in residential conveyancing where there are more than two sellers and/or buyers dependent on each other for completion of their sale and/or purchase.

chairman   Most senior person within a company who presides at meetings of directors where a chairman is often necessary by law.

chairman's report   An annual report on the activities of the company signed by the chairman. This is often given at the annual general meeting and included in the annual report of the company.

Chamber of Commerce   Within the UK voluntary organisations of commercial, industrial and trading members who represent their joint interests to local and central government. Those who belong to the Association of British Chambers of Commerce must abide by its rules and apply for recognition as an incorporated body under one of the Companies Acts.

Chamber of Trade   Local organisation of retailers set up to protect interests at a local level. Most are affiliated to the National Chamber of Trade.

Chancery   One of the three divisions of the High Court.

Chancery Division   One of the three divisions of the High Court of Justice consisting of the Lord Chancellor, a Vice Chancellor and other puisne judges. The work of the division includes the administration of estates, partnership actions, actions relating to mortgages, portions and charges on land trusts, etc. Bankruptcy has also been assigned to the division.

change of use, material (See use classes order.)

CHAR   National campaign for single homeless people formerly known as Campaign for the Homeless and Rootless.

charge   A form of security on land, by virtue of the Law of Property Act 1925, for payment of a debt, allowing the creditor to receive payment from the proceeds of sale. Under the Act the only property charges capable of subsisting are a rent charge or legal mortgage.

charging order A decree of the court imposing a charge on a debtor's property to secure payment of same, made under the Charging Orders Act 1979, as amended.

charitable housing associations Housing associations are normally charities which are either constituted as charitable trusts or companies or industrial and provident societies.

charitable trust A trust by the terms of which the income is to be applied exclusively for purposes of a charitable nature. Trusts of this kind were originally named in the Preamble to the Statute of Charitable Uses 1601. Now see Income Tax Special Purposes Commissions v Pemsel (1891) per Lord Macnaughton: Trusts for the (1) relief of poverty (2) advancement of education (3) advancement of religion and (4) other purposes beneficial to the community not falling under the above heads.

charity Any institution, corporate or not, which is established for charitable purposes and is subject to the control of the High Court in the exercise of the court's jurisdiction, see Charities Act, 1960, Section 45, and Charities Act, 1993, Section 96.

Charity Commissioners A statutory body of Commissioners for England and Wales appointed by the Secretary of State. The body administers charities, secures the effective use of charity property and investigates abuses.

charter mark An award granted by the Citizens' Charter scheme for excellence in the delivery of a public service.

Chartered Accountant Qualified member of the Institute of Chartered Accountants in England and Wales, the Institute of Chartered Accountants of Scotland or the Institute of Chartered Accountants in Ireland.

Chartered Institute of Housing The professional body for those engaged in the practice of housing management, previously called the Institute of Housing Managers.

Chartered Surveyor A professional surveyor who is a fellow or professional member of the Royal Institution of Chartered Surveyors. The Institution is divided into various divisions to which particular members belong.

chattels A term used to describe property other than freehold land, i.e. personal property. Chattels are divided into 'chattels real' and 'chattels personal'. Chattels real are interests less than a freehold in land as they are interests in real estate they are called chattels real to distinguish them from moveables which are called chattels personal. See the Administration of Estates Act 1925, Section 5, subsection 1.

cheap money   Situation where credit is easily obtainable and interest rates are relatively low. A policy sometimes pursued by governments to encourage an expansion in the level of economic activity by reducing the costs of borrowing and investment.

cheque   Pre-printed form on which instructions are given to an account holder, i.e. a bank or building society to pay a stated sum to a named recipient. It is the most common form of payment of debts. An open cheque is payable on presentation by the holder at the drawee's premises. A crossed cheque can only be paid into the bank account of the person presenting it for payment.

chief executive   Person with overall responsibility for the day-to-day running of an organisation.

child en ventre sa mere   Literally a child in the mother's womb. Used in connection with establishing lives in being when determining whether the rules against perpetuity have been complied with.

children's home   A home established under the Children Act 1989, as amended, which provides care for more than three children at any one time. It excludes circumstances where there is some parental responsibility.

chose   A thing. There are two kinds, choses in action which are legally enforceable rights, and choses in possession which are things that can be owned, a moveable chattel; e.g. one's goods.

chose in action   An intangible right which can be enforced by action; such as debt.

chose in possession   A moveable possession.

CIEH   Abbreviation for Chartered Institute of Environmental Health.

CIH   Abbreviation for Chartered Institute of Housing.

CIPFA   Abbreviation for Chartered Institute of Public Finance and Accountants.

circulating capital (working capital)   That part of the capital of a company or other organisation that is used in the activities of the organisation as distinct from its fixed capital or capital assets.

circulation ratio   The relationship between net internal area and gross internal area of a building expressed in terms of a ratio or percentage.

Citizens' Charter Government scheme to promote quality in public services. Under the scheme charter marks are awarded to service providers.

City Challenge Government initiative which aimed to fund regeneration schemes on a competitive basis. This was superseded by the single regeneration budget.

Civic Trust A voluntary organisation established in 1959 to promote the improvement and protection of the built environment.

civil As opposed to criminal, military or ecclesiastical.

claim The assertion of right such as when an event insured against happens.

claimant The party who issues proceedings in litigation.

class closing rules Rules of construction based on Andrews v Partington (1791) relating to gifts to a class. A class will be closed artificially when one member is entitled to be paid having attained a vested interest, thereby excluding all others who have not yet obtained any interest.

class gift A gift is said to be for a class of persons when it is to all those who shall come within a certain category or description. Pearkes v Moseley (1880).

claw back The recovery, by lawful means, of the whole or part of the payment which is properly due at a time stated.

Clayton's case (See appropriation.)

clean hands An equitable maxim; he who comes into equity must come with clean hands. In other words, the plaintiff must have a clear conscience as regards the past. He must be free from reproach or taint of fraud in his conduct in respect of the subject matter of his claim.

clear days Full days within a contract period. Full days do not include the days on which a contract period starts or finishes.

clearance area An area designated by a local authority for the clearance of buildings, where housing especially has been declared unfit for human habitation.

cleared site area Under the Leasehold Reform Act 1967, as amended, a method of valuation using the direct comparison method. Comparison is made with other residential sites which have already been disposed of. The Leasehold Reform Act 1967 allows tenants on long leases the right to acquire the freehold.

clearing bank A member of the bankers' clearing house which facilitates settling of balances arising from cheque and cash transactions. The term is often used as a description for the major high street or joint stock banks.

clerk of works The architect's representative based on site to ensure that the building is constructed in accordance with the contract documents.

client (1) Person who employs the agent. (2) Person or organisation requiring a building to be designed and constructed, and who appoints an architect to organise the project in accordance with his requirements. Some organisations such as local and public authorities employ their own architects.

client/contractor split Internal division of responsibilities within a local authority enabling the authority to deal with requirements of compulsory competitive tendering.

closed economy An economy which neither imports nor exports. This is a theoretical model as no such economy exists in reality.

closed shop An organisation within which an agreement has been reached between the employer and a trade union that only members of that union will be employed by the organisation.

closing Conclusion of a sale where the title of the property is transferred to the new owners and funds are transferred to the appropriate parties (seller, old lender, etc.).

closing order An order made under the Housing Act 1985, as amended, in respect of unfit dwelling houses which are beyond reasonable repair. The order prohibits the use of premises for any purpose other than stated by the local authority.

closing price The price at which shares are quoted on a commodity market or stock exchange at the end of any day's dealings.

closing statement Statement prepared for the buyer and seller itemising all the costs of a transaction.

cloud on title An outstanding claim or encumbrance that, if valid, adversely affects the marketability of title.

club A voluntary association of people meeting together for recreation or social purposes.

cluster flats   Homes in which a number of individuals share facilities such as kitchens and bathrooms, but have their own bedrooms. The accommodation is not usually provided with warden services.

CML   Council of Mortgage Lenders. An organisation which represents the interests of mortgage lenders such as building societies.

code of conduct   A document setting out the guidelines regarding ethical principles and acceptable behaviour expected within a professional organisation or company.

code of measuring practice   A code developed for the measurement of buildings by the Royal Institution of Chartered Surveyors.

codicil   A document that amends a will and must be drawn up in the same way as a will. Often used when the testator wants to add, explain, alter or retract anything.

coding notice   Document issued by the Inland Revenue showing the amount of income tax allowances to be set against an individual's taxable pay.

coemption   The buying up of the whole stock of a commodity.

cohabitation   Living together as husband and wife, even if not married.

cold calling   A system of selling products or services through direct and unsolicited approaches to third parties by mail, door to door canvassing or by telephone.

collar   An agreement made to set a maximum and minimum rate of interest on a variable interest loan or investment in order to limit the risk attached to variable interest rates.

collateral   The security given by the borrower to the principal or lender.

collateral agreement   An agreement used by professionals employed by a developer which places an additional duty of care on whoever purchases the development when completed.

collective bargaining   Bargaining between employers and employees concerning salary, terms of employment, etc.

collective enfranchisement   Provision under the Leasehold Reform Housing and Urban Development Act 1993 allowing leaseholders in blocks of flats to acquire the freehold.

commercial court   A special court of the Queen's Bench Division for dealing with commercial actions.

commercial property   Property intended for use by all types of retail and wholesale stores, office buildings, hotels and service establishments.

commission   The fee of a professional adviser or agent for services provided. Typically, estate agents and mortgage brokers receive a commission for the services they provide. The estate agent secures a buyer for a property that is for sale and a mortgage broker secures a mortgage loan for the buyer to finance the purchase of a property. Commissions are generally paid as a percentage of the sales price in a real estate transaction or the loan amount in a mortgage transaction.

Commissions for Local Administration   Organisations established to investigate allegations of maladministration by local authorities.

committed costs   Costs which an organisation has a long-term responsibility to pay.

committee of management of a housing association   The managing body of a housing association registered as an industrial and provident society and elected by its shareholders. It has ultimate responsibility for the affairs of the association. It must have a minimum of seven members.

commodity   A raw material traded on a commodity market.

commodity broker   A broker who deals in commodities. The rules governing procedures to be adopted vary from commodity to commodity.

common allocations policy   Scheme within which social landlords within a local authority area agree to operate the same criteria for making lettings. This arrangement will usually involve the operation of a common waiting list.

common assurances   Legal evidence to prove transfer of property, e.g. by deed, will, etc.

common housing register   Name for common waiting list following the Housing Act 1996.

common law   (1) The law based on 'the common sense of the community' developed by the Royal Court between the 11th and 13th century to apply to the whole nation and supplant local practice. (2) Law embodied in case precedent as opposed to statute law or equity.

common law wife or husband   A man or woman who live together but are not married.

common parts The parts of the building not let to individual tenants and normally retained by the landlord, usually applied to parts of multi-occupied buildings.

commonhold A situation where the freehold of all common parts of a property is vested with all the owners and managed by a separate management committee. Commonly used by flat owners.

commorientes Persons dying together at the same time are presumed to have died in order of seniority.

community based housing association A housing association with a high level of tenant involvement.

community care The provision of services and support within the home to vulnerable people in order that they may live as independently as possible.

community charge A local tax replaced by council tax.

community leasehold A form of equity sharing through housing associations. It is not the same as shared ownership in that occupiers do not have the right to acquire increasing shares in the equity of their homes. Rents are based on capital values and not fair rents.

community ownership The process where a housing association or co-operative buys out a local authority estate.

community property Property commonly owned between husband and wife.

Companies House Location of the Registrar of Companies, the statutory authority responsible for the incorporation of limited companies.

company A corporate body created by Royal Charter, Act of Parliament or registered with the Registrar of Companies under the Companies Act 1985. A registered company is the most common and it may be registered either as a public limited company or a private company.

company formation Procedure for forming a company within the UK. Subscribers to the company must send the Registrar of Companies documentation including names and addresses of directors and secretary, a declaration of compliance (to the provisions of the Companies Act), the Memorandum of Association and the Articles of Association.

company limited by guarantee A company with a constitution under which the liability of members is limited by the Memorandum of Association to amounts that they have agreed to undertake to contribute in the event of winding up.

company limited by shares A company within which the liability of members is limited by the Memorandum of Association to the amounts paid or due to be paid for shares. This is the most usual form of company within the UK.

company registration The requirement by the Registrar of Companies for certain types of companies to register and receive a Certificate of Incorporation recognising its legal identity.

company seal Seal engraved with company's name which is used to authenticate share certificates and other documents of importance issued by the company. The Articles of Association of the company will set out how and when the seal is to be fixed.

company secretary An officer of the company concerned with keeping the company's statutory books and supervising the administration of its affairs. Legal duties include the submission of the annual return, keeping of the minute books for board and company meetings, maintenance of the share register and payment of the dividends and interest.

comparables Properties that are similar in value to a particular property and are used as comparisons to determine the fair market value of a specified property.

compensation Money paid under the Land Compensation Acts 1961 and 1973, as amended, to the owner and/or occupier of property in compensation for curtailment or removal of their rights in a property, often awarded in association with compulsory purchase.

competition Rivalry between suppliers of goods and services within a market.

competitive market Market where goods and services are freely and voluntarily offered for sale by any number of willing sellers to any number of willing buyers at prices agreed by both parties.

completely constituted trust This is a trust which has been perfectly created and the property has been fully and finally vested in the trustee to be held for the benefit of the intended beneficiaries, or a settler has declared himself a trustee of that property.

completion   Completion of a contract on the part of the vendor in conveying title and acceptance of same by purchaser.

completion statement   A statement prepared by solicitors normally for vendor and purchaser respectively, which is made following completion of the conveyance of a property.

compound interest   Interest made at given intervals on accumulated interest and including the principal.

compulsory competitive tendering   Requirement upon local authorities to put out to tender work previously carried out by council staff.

compulsory purchase   The taking of private property for public use by a government unit, against the will of the owner, but with payment of just compensation under the government's power of eminent domain.

compulsory purchase order   An instrument facilitating the compulsory acquisition of land. Such instruments can only be issued for purposes specified in particular enabling acts.

compulsory winding up   The winding up of a company by the court following presentation of a petition at both the court and the registered offices of the company.

concurrent interest   Two or more interests with joint ownership of land.

condition   Circumstances essential to the occurrence of an event. (See condition precedent and condition subsequent.) An important term of a contract.

condition precedent   A condition which delays the vesting of a right until the occurrence of a particular event, e.g. to 'Z' if she graduates as an accountant.

condition subsequent   A condition which provides for the defeat of an interest on the occurrence or non-occurrence of a particular event, e.g. to 'Z' if she always lives in England.

conditional commitment   A lender's promise to issue a loan subject to certain conditions. Generally, the lender will not fund the loan until the conditions have been met.

conditional contract   A contractual agreement which is subject to either a condition subsequent, or a condition precedent, or both.

conditional interest   An interest on condition subsequent. (See condition.)

conditional legacy A bequest in a will whose operation depends on the happening or not happening of some uncertain event on which it is either to take effect or to be defeated.

conditional offer Purchase offer in which the buyer proposes to purchase property only after certain events (sale of another home, finding a loan commitment, etc.) occur.

conditional planning permission A colloquial term describing the grant of planning permission with conditions attached.

conditional will A will executed with the intention that it shall be rendered operative only on the occurrence of a specific event.

conditions of sale The conditions attaching to the sale of properties under offer. Two standards exist; firstly the National Conditions of Sale, and secondly the Law Society's Conditions of Sale.

condominium A development including two or more units, normally residential. The individual units, or parts thereof, being owned, but other common parts managed separately by an agreement of the owners.

Confederation of British Industry (CBI) Organisation formed in 1965 by a merger of the National Association of British Manufacturers, the British Employers' Confederation and the Federation of British Industry. It is a politically neutral organisation which represents industry in dealings with government. Its governing body is the Confederation of British Industry Council and there are 30 regional councils which deal with local industrial issues.

confirmation The conveyance of an estate, whereby a voidable estate is made sure.

conglomerate Group of companies usually managed by a single holding company. The group may be characterised by considerable diversity with little integration and few transactions between each of the subsidiaries.

consent Colloquial term used to describe when planning permission has been granted for development.

conservation area An area designated under the Planning (Listed Buildings and Conservation Areas) Act 1990, as amended, by the local planning authority as being of special architectural and/or historic interest.

consideration (1) Anything of value given to induce another to enter into a contract. (2) The sum payable by one party as part of a contractual agreement. Regarding property law, the term is normally applied to the price offered by the purchaser in a contract to acquire land or buildings. (3) An exchange of promises between parties making a contract by which one party buys the promise of the other. Consideration is executed when the act constituting the consideration is performed and it is said to be executory when it is in the form of promises to be performed at a future date.

consistency concept   Concept used in accountancy to ensure consistency of treatment of similar items within each accounting period and from one period to the next. It is a principle contained in the Statement of Standard Accounting Practice.

consolidated accounts   A company which has subsidiary companies in addition to the normal requirements of the Companies Act must file group accounts. These are intended to show the state of affairs of the group at the end of a relevant financial period. The most common form of consolidated accounts involves the combination of individual, final and audited accounts of the various companies into one set of accounts where inter-company transactions and indebtedness are eliminated on a self-cancelling basis and the end product consists of a consolidated balance sheet and a consolidated profit and loss account.

consolidated balance sheet   The balance sheet of a group which combines the individual final accounts of the various companies within the group.

consolidation   Joining of two separate actions so that they can be tried together.

consolidation (of capital)   An increase in the price of a company's shares brought about by combining a number of lower priced shares into one higher priced share, e.g. five £2 shares may be consolidated into one £10 share. This will usually be done by an ordinary resolution at a general meeting of the company.

consolidation of mortgages   The doctrine which allows a mortgagee with several mortgages by the same mortgagor on several properties to redeem all of them if the mortgagor seeks to redeem any of them. The doctrine is now repealed under the Law of Property Act 1925.

consols   Government consolidated stock which is irredeemable. These government securities pay interest but have no redemption date.

consortium   A combination of two or more parties coming together for a project or series of projects without losing their own identity.

A consortium is often formed for the purpose of bringing together a range of skills or to eliminate competition.

construction (1) The process of ascertaining the meaning of a written document. (2) The judicial interpretation of statutes.

construction loan Short-term financing for construction.

construction period The length of time taken from the start of construction to completion of a particular building project. The completion date is often referred to as the date of practical completion and a certificate is issued by an architect to that effect.

constructive Not directly expressed, inferred, implied.

constructive trust A trust imposed by equity regardless of the party's intentions in order to satisfy the demands of justice and conscience.

consumer credit licence Under the Consumer Credit Act 1974, as amended, a housing association requires a licence if it wishes to offer credit brokerage services introducing tenants to sources of mortgage finance.

Consumers' Association Independent non-profit making organisation which was set up in 1956 to help consumers by testing goods and services. The Consumers' Association publishes the monthly magazine *Which?* and various paperback publications.

contentious business Business before a court or an arbitrator not being business which falls within the definition of non-contentious or common form probate business in the Supreme Court Act 1981, Section 128.

contiguous A common boundary.

contingency (1) Something related to a possible future and uncertain event. (2) A condition that must be satisfied before the buyer can consummate the purchase of a property. Contingencies are generally outlined in the purchase contract between the buyer and seller.

contingency fee An incentive fee which is paid in relation to the degree of success achieved in the task at hand.

contingency insurance In an insurance contract, the protection offered on the occurrence of a risk, or other event, insured against.

contingent interest An interest in land which comes into operation at some time in the future, often associated with a specific event.

contingent legacy One bequeathed on a contingency, e.g. if the legatee shall attain the age of 30 years.

contingent liabilities Accounting term for liabilities that may or may not arise in the future.

contingent remainder A remainder is contingent if the person to receive it, the grantee, is unascertained or if the title depends on the occurrence of a designate, for example a grant to 'Y' for life, remainder to his first daughter to qualify as a lawyer.

contract An agreement between two or more parties which is legally enforceable. For a legally enforceable contract to exist there must be: (1) legal capacity of the parties to contract; (2) intention to contract; (3) valuable consideration; (4) legality of purpose; and (5) certainty of terms.

contract deposit The amount paid by the purchaser of property under a contract and usually held by a solicitor. It indicates that the payer will fill his contractual obligations, but in the event of a default, he will forfeit the deposit.

contract documents The documents which form the basis of a contract between the client and contractor and which include a formal signed contract by both parties and the design information (drawings, specifications and bill of quantities). (See also contract sum; contract period.)

contract for sale of land A written agreement signed by both parties. Where such agreements are exchanged, one part must be signed by each party under Section 2 of the Law of Property Act 1925, as amended.

contract manager A person employed by a contractor and who is responsible for managing the building work.

contract of purchase An agreement between parties for the sale of an estate. In some cases, it is synonymous with a purchase agreement, sales agreement, or land contract.

contract of sale A purchase transaction in which the buyer receives possession of the property, but the seller retains title.

contract period The time agreed by the contractor to complete the building in accordance with the contract documents commencing on a specified date (i.e. date of commencement) and completing on a specified date (i.e. date of completion).

contract sales price The full purchase price as stated in the contract.

contract sum   The sum of money agreed by the contractor to carry out the building works in accordance with the contract documents.

contractor   A building firm who are responsible for constructing a building in accordance with the contract documents.

contractor's quantity surveyor   A person who is responsible for dealing with all contractual and financial matters on behalf of the contractor, and would prepare monthly and final accounts for agreement by the design team's quantity surveyor. (See quantity surveyor.)

contractual improvement   The improvement of property carried out under a legally enforceable agreement.

contribution   The difference between sales value and the variable cost of a sale, i.e. the amount that a given transaction provides to cover fixed overheads and provide profit.

contribution clause   The clause in an insurance policy which requires part payment of the losses suffered.

contributory mortgage   Where the mortgage advance is made to two or more parties. A trustee is precluded from receiving such an advance.

contributory pension   Pension in which the employee as well as the employer contributes to the pension fund.

control process   Process which helps the organisation to determine whether or not its objectives are being achieved.

controllable costs   Costs identified as being controllable and so potentially influenced by activities of management.

controlled tenancy   A dwelling house which has a protected or statutory tenancy where the rateable value did not exceed that stated in Section 17(1)(a) of the Rent Act 1977. Controlled tenancies were converted to regulated tenancies under Section 18A of the Rent Act 1977, as amended.

conversion costs   Costs incurred in converting raw material into finished goods. These costs will usually include direct labour and production overheads but exclude the cost of raw materials.

conversion   In equity this is the notional change of land into money or money into land. The effect is to turn realty into personalty and personalty into realty for all persons.

conveyance   The transfer of land under the Law of Property Act 1925, as amended.

conveyancer   A solicitor who specialises in conveyancing.

conveyancer, licensed   One authorised under the Administration of Justice Act 1985, as amended, to undertake conveyancing for profit.

conveyancing   The legal procedures involved in the transfer of title from one person to another.

conveyancing ombudsman   One who investigates complaints against authorised practitioners. Established under the Courts and Legal Services Act 1990, as amended.

co-operative housing   A building or group of dwellings owned by a corporation, the stockholders of which are the residents of the dwellings. It is operated for their benefit by their elected board of directors. In a co-operative, the corporation or association owns title to the real estate. A resident purchases stock in the corporation that entitles him to occupy a unit in the building or property owned by the co-operative. While the resident does not own the unit, he has an absolute right to occupy his unit for as long as he owns the stock.

co-operative promotion allowance   Housing association grant allowance to pay certain costs when a primary co-operative receives services from another organisation, usually a secondary co-operative.

co-ownership   The situation where two or more people are entitled to the shared ownership of land, either by a joint tenancy or by tenancy in common.

core and cluster   System of supported housing where people requiring low levels of support live independently but are linked to a centrally based staff team.

cornering the market   Obtaining a virtual monopoly over the supply of a good or service. This possibility is now rare because of government restriction on monopolies.

corporate culture   The values, beliefs, norms and traditions which influence the behaviour of people within an organisation.

corporate finance   The funding of businesses.

corporate governance   The way in which organisations are managed and the nature of accountability of the managers to the owners. The Cadbury Report (1992) set out guidance in a code of practice.

corporate plan   (See business plan.)

corporation An association of persons recognised by law as having rights and liabilities distinct from the individuals forming the corporation. A corporation is either a corporation sole or a corporation aggregate.

corporation aggregate A corporation composed of more than one individual. The most common forms of corporation aggregate are local councils and limited companies.

corporation sole A corporation consisting of one individual normally held in some public office under the Crown or the state, e.g. a bishop or the sovereign.

corporation tax Tax payable on the total profits of a UK company within each accounting period. The rate of corporation tax depends on the level of profit of the company.

corporeal property Land.

cost accounting Division of management accounting concerning the everyday running of a business which includes collecting, processing and presenting financial information within an organisation.

cost basis Accounting figure that includes original cost of property plus certain expenses to purchase, money spent on permanent improvements and other costs, minus any depreciation claimed on tax returns over the years.

cost benefit analysis Technique used to calculate the social and economic worth of a prospective project. It attempts to attach monetary value to both calculable gains and losses of a project and also the abstract benefits or possible negative results.

cost centre Area of an organisation for which costs are ascertained. Cost centres may be based on functions, departments or individuals and are generally categorised as production cost centres or service cost centres.

cost convention Custom used as the basis for recording costs against profit for an accounting period. The cost convention may be based on historical cost, current cost or replacement cost.

cost floor Under the right to buy scheme the original cost of provision of the property. The discount allowable cannot fall below the cost floor.

cost of living clause The clause in a lease or other form of property contract providing for the adjustment in rent price, or other financial item, based upon the retail price index.

cost plus contract   A building contract where the price is based upon the actual, or estimated, cost of the works, together with a proportion to represent profit for the contractor.

cost plus pricing   A method of pricing in which the cost of one unit is calculated and then a percentage mark-up is added.

cost rent   A rental payment calculated to provide, over a period of time, a sum which is sufficient to meet the cost over the same time period of a notional loan interest, or actual interest.

cost unit   A quantitative unit of produce or service in relation to which costs are ascertained.

council tax   Local property tax which replaced the community charge levied by local councils.

counterclaim   A cross action brought by the defendant against the claimant, or by the respondent against the claimant.

counteroffer   A new offer made because of another offer, which cancels the original offer.

counterpart   A duplicate document of a legal nature often associated with the landlord's copy of a lease.

Countryside Act 1968   This Act replaced the National Parks Commission with a countryside commission, which was given wider powers and improved finance. Local authority power was also expanded to include the provision of country parks, places intended for enjoyment.

county courts   Local courts which deal with smaller civil actions.

court of appeal   The court which hears appeals from the High Court and County Courts (amongst others).

court of protection   The court which administers the property of mentally disordered persons within the meaning of the Mental Health Act 1983.

covenant   (1) An agreement written into deeds and other instruments promising performance (or non-performance) of certain acts or stipulating certain uses (or non-uses) of the property. It may be positive or negative. (2) An agreement contained in a contract or a deed whereby a party stipulates the truth of certain facts or binds himself to give something to another or to do or not to do any act.

covenants for title   The agreements entered into by the vendor giving the purchaser the right of action if the title proves bad under the Law of Property Act 1925, as amended.

credit (1) The sum which a supplier allows a customer before requiring payment. (2) In double-entry book keeping an entry on the right-hand side of an account showing a positive asset.

credit line Facility for borrowing over a given period to a specified extent.

credit rating Assessment of the credit worthiness of an organisation or an individual suggesting the extent to which they can be granted credit.

creditors Person or entity to whom money is owed.

cremation The disposal of a dead body by burning in a crematorium. See the Cremation Acts 1902 and 1952.

Criminal Law Act 1977 An Act which makes it a criminal offence to use or threaten violence to secure entry to property if there is someone on the premises who is opposed to entry.

crown court The branch of the Supreme Court which deals with criminal trials and some civil cases.

cumulative legacy A legacy which is to take effect in addition to another disposition in favour of the same party as opposed to a substitutional legacy which is to take effect as a substitute for some other disposition.

current account An active account opened by an individual or an organisation with a bank or building society. An account into which monies are paid or from which sums are withdrawn either in cash or by cheque in the course of everyday affairs.

current assets Accounting term for assets which form part of the circulating capital of a business and are turned over frequently in the course of trade. These will include cash, debtors, stock and work in progress.

current cost accounting System of accounting based upon a concept of capital in which net operating assets such as fixed assets, stocks and working capital are expressed at current price levels.

current liabilities Those liabilities which an organisation would usually expect to settle within a relatively short period of time. These will include trade creditors, dividends, short-term loans and tax due for payment.

51

current ratio   One of two common liquidity ratios. Current ratio is an expression of the ratio of current assets to current liabilities, e.g. if current assets are £50,000 and current liabilities £25,000 the current ratio is 2:1.

custodian trustee   Office created by the Public Trustee Act 1906, a trustee appointed to have the custody as distinguished from the management of the trust property. Among those who may act as custodian trustee are the Treasury Solicitor and trust corporations.

customer service   Goods and services an organisation offers to its customers.

cy-pres doctrine   A charitable trust which by its terms is impossible initially or is impractical or becomes so subsequently, will not necessarily fail. The court may apply the trust property cy-pres by means of a scheme to some other charitable purpose which resembles the original purpose as nearly as possible. See Charities Act 1993.

D

damages   A monetary sum awarded by a court to a victim of breach of contract to compensate a loss or injury, They can be classified as nominal damages, which may be awarded if no actual damage has been caused, substantial damages may be awarded when there has been actual damage, and exemplary damages may be awarded to punish the defendant, liquidated damages being pre-determined in an agreement and unliquidated damages which are subject to the discretion of the court.

damping formula   A procedure used by the Department of the Environment to slow the effects of the change to housing needs indicator scores.

dangerous premises   Under the Occupiers Liability Act 1957, as amended, the occupier owes a duty of care to lawful visitors. Compensation for loss to the visitor may be payable by the occupier and not the owner. This doctrine extends to trespassers.

dangerous structure notice   A notice, issued by the local authority, to a building owner requiring him to put a building into a safe condition.

data Information which is processed, stored or produced by a computer.

data protection Safeguard to prevent misuse of personal data about individuals which is stored on a computer. The principles of data protection, responsibilities of data users and rights of individuals are governed by the provisions of Data Protection Act (1984).

date of valuation A specific date stated on a valuation certificate which indicates the value of the building on that particular date.

day book Book or journal of prime entry recording purchases and sales as they take place or containing lists of invoices in and invoices out. Day book entries are transferred to ledgers.

daylight factor Used to calculate the amount of glazed area required to provide the required degree of natural illumination for a room. (See natural light.)

daywork contract A small works building contract where payment is based on hours worked, cost of materials, plant, transport and a percentage for the contractor's overheads and profit.

de ingressu A writ of entry to property.

death The cessation of life processes and all vital signs, not defined by statute.

death duties Estate duty paid on property which passes at death.

debenture Most common form of long-term loan taken by a company usually repayable at a fixed date with a fixed rate of interest. The interest must be paid before a dividend is paid to shareholders. Most usually debentures are secured on the borrower's assets with the exception of naked or unsecured loan stock. The term also refers to a deed under seal setting out the main terms and conditions of such a loan or to a form of bank security covering corporate debt where the bank ranks as a preferred creditor in the event of liquidation.

debit An entry on the left-hand side of an account in double-entry book keeping representing an increase in the organisation's assets or a decrease in its liabilities.

debit balance The balance of an account where total debit entries exceed the total credit entries.

debt Any amount one person owes to another.

debt collecting agency Organisation concerned with the collection of debts on behalf of other people.

debt profile Often contained within an organisation's business plan, a profile showing the overall impact of an organisation's borrowing over the full length of repayment period of all loans.

debt rescheduling Negotiated rescheduling of debt to take the form of an entirely new loan or an extension of the existing loan repayment period.

debtor A person or entity who owes money to an individual or an organisation. Usually referring to customers who have yet to pay for goods and services already provided.

declaration of trust A declaration whereby a person admits that he or she holds property on trust for another. If the declaration is of a trust of land it must be evidenced in writing signed by the party declaring the trust; see Section 53, Law of Property Act 1925. For declarations of trusts of money or personal chattels no formalities are necessary.

deduction of title Expression used to signify the seller's obligation to prove to the buyer his ownership of an interest in land which he is purporting to sell. Ownership is proved to the buyer by producing documentary evidence of title.

deed A written document that must be signed by the person making it in the presence of witnesses. For example, a deed transferring title to land. The deed must contain an accurate description of the property being conveyed, be signed and witnessed and be delivered to the purchaser at closing day.

deed of covenant Covenants or contractual promises are often entered into by separate deed.

deed of postponement Arrangement where the lending authority postpones its right to first claim on a housing association's funds in favour of another lender.

deed of trust A document that gives a lender the right to foreclose on a piece of property if the borrower defaults on the loan. An instrument given by the borrower to a third party (trustee) vesting title to the property in the trustee as security for the borrower's repayment of the mortgage loan.

deed restriction Restrictions placed on use of land by writing in a deed to control use and occupancy of the property by future owners.

deemed planning permission The situation under the Town and Country Planning Acts where planning permission is considered to have been granted without the necessity of formal approval from the local planning authority. (See general development order.)

default   The failure to do something required by the law. For example, one party not meeting the requirements of a specific court order or the failure to make mortgage payments or violations of other provisions of the mortgage note.

defeasance   The ending of an interest in property in accordance with conditions stipulated in a separate instrument or document.

defective title   Title to real property which lacks some of the elements necessary to transfer good title. Title to a negotiable instrument obtained by fraud.

defects liability period   The period agreed following the practical completion of building engineering, or other construction works, during which the contractor is obliged to make good any failure of materials or workmanship to meet the specific terms of the contract.

defence   A pleading from the defendant in answer to the statement of claim.

defendant   The person sued in an ordinary civil action.

deferred interest mortgage   A mortgage when interest payment is delayed until well into the mortgage repayment period.

deferred maintenance   Any repair or maintenance of a piece of property that has been postponed, resulting in a decline in property value.

deforcement   Wrongful holding of lands of another.

dehors   Beyond; separate; unconnected with.

de-layering   The reduction of the number of layers within an organisation usually involving the removal of layers of middle management from organisational structures in order to provide a flexible and responsive organisation.

delegation   The assignment of a duty or power of action to another person with sufficient authority to act for the person assigning the authority.

delegatus non potest delegare   Literally an agent cannot delegate his authority. A trustee may appoint an agent subject to the terms of the trust instrument. See currently the Trustee Act 1925, Sections 23 and 25, and the Trustee Act 2000.

delinquent mortgage A mortgage that involves a borrower who is behind on payments. If the borrower cannot bring the payments up to date within a specified number of days, the lender may begin foreclosure proceedings.

demand curves Graphical representation showing the relationship between the quantity of a good demanded to its price. In normal circumstances the demand curve will slope downwards suggesting that an increase in the price of a good results in a lower level of demand.

demand pull inflation A rise in prices caused by an excess of demand over supply in the economy.

demise The granting of a lease or the term of years granted.

demolition order An order of a local housing authority requiring the pulling down of a dwelling house which is unfit for human habitation, and which cannot be repaired at reasonable expense.

demonstrative legacy A gift by will of a certain sum to be paid out of a specified fund or specified part of the testator's property, e.g. £500 out of my current account with National Westminster Bank.

density A term often used in town planning, meaning the number of dwelling units per area of land, often expressed as dwellings per acre or hectare.

deposit Money given by the buyer with an offer to purchase property. The deposit is money given to the seller or his agent by the potential buyer upon the signing of the agreement of sale to show that he is serious about buying the house. If the sale goes through, the money is applied against the down payment. If the sale does not go through, the deposit may be forfeited to the seller.

deposit account An account with a bank from which money cannot be withdrawn by cheque. Interest paid will be dependent on the current rate of interest and the period of notice required by the bank before money can be withdrawn.

deposit of title deeds Delivery of same to creditor as security for debt, often a mortgage.

depreciation The internal charge made by an organisation against its revenue to provide for the progressive deterioration or obsolescence of its fixed assets. Provision for depreciation may be computed by a number of methods. Depreciation reduces book value of the asset and is charged against the income of an organisation.

de-regulation The removal of controls imposed by governments on the operation of markets.

derelict Buildings which have been abandoned or neglected.

descendant A person descended from an ancestor, generally refers to lineal descendant only.

descent The passage of an interest in land upon the death intestate of the owner to a person or persons by virtue of consanguinity with the deceased.

design and build A project in which the owner contracts directly with an individual or company to perform design and construction.

design and build contract A form of contractual agreement for the undertaking of building, engineering, or construction works which incorporates both the design and production phases of the operation.

design guide A document, often associated with town planning, which offers guidance as to the type of buildings, or alterations to buildings, which would be acceptable within a defined area.

design team The people responsible for designing a building and producing the design information (drawings, specifications, bill of quantities) and co-ordinated by the architect, including such persons as quantity surveyors and engineers. (See architect, quantity surveyor, engineer.)

detached house A dwelling that is complete and is not joined to others. (See semi-detached house, terraced house.)

detailed drawing (See drawings.)

determinable interest These are interests which are terminable or end on the happening of a specified contingency.

determine To come or to bring to an end.

detinue An action in tort for recovery of a specific chattel.

devastavit Literally he has wasted. A personal representative who mis-applies or mismanages the assets of a deceased person is answerable for that waste which is said to constitute a devastavit.

developer A person who develops land or property for profit.

developer's profit The portion of the value of a property accruing after allowing for the acquisition, construction costs, and other outlays of a developer.

developer's risk and profit   Often associated with the residual value method of property valuation being the amount which is allowed to cover firstly, an estimate of the sum needed to reflect the risk element between the valuation date and the completion of the development programme, and secondly, an amount to meet the developer's requirements for profit on the venture.

development agent   In terms of the housing association movement a housing association or a secondary co-operative which furnishes development services for smaller associations.

development brief   A document which contains a statement by the owner of a site suggesting detailed requirements for the proposed development to would-be developers.

development corporation   A corporate body usually a public authority charged with the development of a specific site, for example, New Town Development Corporation.

development expenditure   The total sums of money expended by a developer in undertaking a development project. Normally divided into the cost of land acquired, and the costs of construction.

development, permitted   Planning permission granted by a development order which does not require formal planning approval from a local planning authority. The Town and Country Planning (General Permitted Development) Order 1995 specifies certain classes of development which may be undertaken without formal permission.

development plan   A plan, usually associated with town planning, which indicates how the local planning authority would like to see a specific area developed in the future.

devise   A gift by will of real property such as an interest in land. The giver is called the devisor and the recipient the devisee. (See bequeath and bequest with reference to personalty.)

devisee   One to whom real property is given by will.

devisor   A testator who leaves real estate.

dilapidations   The items of disrepair arising through breach of a contract, especially by one of the parties to a lease giving rise to a claim for remedial action or damages.

dilapidations report   A list or schedule of remedial work required to be carried out to an existing building, together with the costs involved, and normally compiled from a building survey.

dimension The distance between two points and included on scale drawings to indicate lengths, widths and heights of various parts of the building.

dimension plans Plans which show the layout of a house but are less detailed than full blueprints.

diminishing balance method Method of calculating the depreciation of a capital asset within an accounting period in which the percentage to be charged against income is based on the depreciated value at the beginning of the period. The percentage is calculated so that the asset will be written down to scrap value after a stated number of years.

direct access accommodation Accommodation provided by way of a hostel for single people who are rootless.

direct comparison method A valuation method whereby the rental or capital value of a property is assessed with regard to the prices or rents recently achieved by similar properties in a similar locality.

direct costs Costs which can be traced directly to units of production. These will include materials, labour and production expenses.

direct debit Form of standing order given to a bank by an account holder to pay regular amounts from a cheque account to a third party. Unlike a standing order the amount to be paid is not specified.

direct labour Labour directly concerned with the production of a product or service.

direct labour organisation (DLO) Department of an organisation which carries out building or maintenance work directly for that organisation.

director Individual charged with responsibility to carry out the day-to-day management of a company. A public company must have at least two directors, a private company at least one.

directors' fees Amounts received by directors including salaries, fees, wages, expenses and benefits paid or provided by the employer which are deemed to be remuneration or emoluments.

directors' report Annual report by the directors of a company to its shareholders. The report forms part of the company's accounts which are required to be filed with the Registrar of Companies.

disbursements Payments made during the course of an escrow or at closing.

disclaimer (1) A refusal to act, for example as the beneficiary under the terms of a will. (2) A move by a person to limit liability attaching to him or her. (3) The renunciation of a claim or right.

disclosure A statement to a potential buyer listing information relevant to a piece of property, such as the presence of radon or lead paint.

discontinuance order An order made by a local planning authority under the Town and Country Planning Act 1990, as amended, requiring the discontinuance of the lawful use of land or buildings. Compensation may be payable.

discount bonds Bonds issued at below their redemption value. Used by the housing association movement.

discounted cash flow An evaluation of the future net cash flows generated by a capital investment project by discounting them to their present-day value.

discounting A statistical procedure by which amounts of money due to be received in the future are brought to their current value on a specific valuation date, allowing for accumulated interest at a selected rate which it is assumed would be earned during the intervening period.

discovery Disclosure of all the documents relating to a case before trial.

discretionary trust A trust under which trustees have an absolute discretion to apply the income and capital of the trust as they wish. No beneficiary is able to claim a right that any part or all of the income is to be paid to him or applied in any way for his benefit.

disposing lender The Law of Property Act 1925 s101 gives a power to sell the legal estate vested in the borrower, subject to prior encumbrances but discharged from subsequent ones, to every lender whose mortgage is made by deed. The power of sale arises when the mortgage money becomes due under the mortgage, i.e. on the legal date for redemption, which is usually set at an early date in the mortgage.

disposition The act or process of transferring something to or providing something for another, see section 53 of the Law of Property Act 1925 which provides that the disposition of any subsisting equitable interest must be in writing, signed by the person disposing of the interest or his or her lawfully authorised agents.

distressed property Property that is in poor physical or financial condition.

distributable profits The profits of a company available for distribution as dividends.

distributable reserves Accumulated retained profits of a company that it may legally distribute by way of dividends.

distribution of estate The apportioning of the estate of a deceased intestate among the persons entitled to share in it.

district valuer A public officer responsible for undertaking valuations for taxation, compulsory purchase and other statutory purposes on behalf of government departments or local authorities.

disturbance Displacement of a person's home following compulsory purchase or possession secured by the landlord. Compensation may be payable to the displaced person.

disturbance payment The amount of compensation payable by a local authority in pursuance of its statutory powers. Often associated with compulsory purchase.

diversification Spreading of an investment portfolio over a wide range of companies or the movement of an organisation into the provision of a wider range of goods and services.

dividend Distribution made to shareholders in proportion to the number of shares that they hold generally from post-tax profits. An interim dividend is paid half yearly, the final dividend at the year end.

dividend cover Number of times a company's dividends to ordinary shareholders could be paid out of its profits after tax in the same period.

divorce The dissolution of a marriage on the grounds of an irretrievable breakdown.

divorce value Additional value released by the subdivision of property into two or more parts.

do it yourself shared ownership (DIYSO) Shared ownership schemes in which a potential purchaser identifies a suitable property which is then acquired by a registered social landlord which sells on to the purchaser on a shared ownership basis.

dogs Term derived from the Boston Matrix to describe goods with a low market share in new or slowly growing markets. These goods are unlikely to yield attractive profits.

domicile Often associated with taxation purposes and being the country which is regarded as an individual's permanent home.

donatio mortis causa A gift of property made by a person who expects to die. The gift must be made in contemplation of death, it must be conditional on death and the property must be capable of being delivered. See Wilkes v Allington, Woodard v Woodard.

donee A person who receives a gift from a donor.

donor A person who makes a gift to another.

double-entry book keeping Method of book keeping such that each transaction is entered twice, e.g. when a debtor pays cash for goods purchased the cash held by the business is increased and the amount due from the debtor is decreased by the same amount. This system enables the business to be controlled because all books of accounts must balance.

double probate A grant of probate made to an executor to whom power has been reserved to prove at a later date all on the happening of a specified event and who has proved – for example, an executor who was an infant and who has later attained his majority.

dower The right of a widow to a life interest in the real property of her deceased husband. Abolished by Law of Property Act 1925 and Administration of Estates Act 1925.

down payment The amount of money a buyer agrees to give the seller when a sales agreement is signed. Complete financing is later secured with a lender. Down payment is the difference between the sales price and the mortgage amount. Buyer cash required at closing includes the down payment, closing costs and prepaid expenses.

down sizing Reduction in the size of an organisation usually by reducing the number of direct employees in order to save costs and increase flexibility.

dowry Within social housing the sum of money which the vendor must pay to the transferee when a property to be transferred from a local authority to a registered social landlord is in such poor condition that it has a negative value once the cost of repairs is taken into account.

draw down The release of finance, often at various stages, for development projects provided by way of a loan often for large sums over a long period of time.

drawings Part of the design information which are dimensional details of the proposed building drawn to scale and include elevations, sections and plans.

dual registration An establishment which is registered both as a residential care home with the local authority and as a nursing home with the district health authority.

duality of interest When an individual who may be serving as an officer or member of a housing association receives payment or benefit from the association beyond his normal contract of employment with the association.

duplicate will A will executed in duplicate, the intention usually being that the testator keeps one copy and the other is to be deposited with someone else. On probate, both copies must be deposited at the probate registry.

durante absentia Literally during absence. Administration is granted *durante absentia* when an executor is out of the realm.

dwelling (1) A building used for residential purposes. (2) Defined under the Rent Act 1977 as 'a house, or part of a house'. Defined under the Housing Act 1985 as 'premises used, or suitable for use, as a separate dwelling'. Under the Landlord and Tenant Act 1985 defined as 'a building or part of a building occupied, or intended to be occupied as a separate dwelling, together with any yard, garden, outhouses and the permanencies belonging to it, or usually enjoyed with it'.

dwelling house Defined under the Housing Act 1985 in reference to secure tenancies as 'a house or part of a house'.

dying without issue These words are held under the Wills Act 1837, section 29, to refer only to the case of a person dying and leaving no issue behind at the date of death. A rule of construction unless a contrary intention shall appear by the will.

E

early occupancy The condition in which buyers can occupy the property before the sale is completed.

earnings per share The profit attributable to each ordinary share in a company based on the net profit for the period divided by the number of equity shares in issue that rank for dividend in respect of that period.

easement A legal right – created by grant, reservation, agreement, prescription or necessary implication – which one has in land owned by another – e.g. Right of Way, a right given to a third party to use a portion of the property for certain purposes, such as power lines or water mains. The land owned by the possessor of the easement is called the dominant tenement, and the servient tenement is the land over which the right is enjoyed. A positive easement consists of a right to do something on the land of another and a negative easement restricts the use the owner of the servient tenement may make of his land.

economic life The time during which the value of buildings or a site, in any particular use, is greater than the value of the site for any other feasible purposes, including redevelopment.

economic obsolescence Impairment of desirability or useful life arising from economic forces, such as changes in optimum land use, legislative enactments which restrict or impair property rights and changes in supply–demand relationships. Loss in the use and value of property arising from the factors of economic obsolescence is to be distinguished from loss in value from physical deterioration and functional obsolescence.

economic rent Rent necessary to cover all costs in the absence of a subsidy. This calculation excludes profit and is often used as a basis of comparison with fair and assured tenancy rents.

elderly persons' dwellings Residential units suitable for active elderly people, designed in accordance with the Housing Corporation's Design Guidelines on Sheltered Housing and Mobility Standards. (See mobility housing.)

elevation A drawing which represents to scale the completed building as viewed from the front, rear and sides. (See drawing.)

eligible rent Term used within housing benefit calculations to indicate the proportion of a rent which pays for the property but excludes items such as water rates.

emoluments Total benefits received by an employee which will include salary and goods and services such as private use of a company car, subsidised accommodation and meals.

employee buy out The purchase either individually or through an employee trust of a controlling interest in a company's equity share capital by the employees.

employers' liability insurance Insurance policy which will cover injury, death or industrial disease incurred by employees whilst on the employers' premises.

employment agency Organisations which exist to bring prospective employers into contact with job seekers. These organisations may be established privately or by the state.

employment protection Safeguarding of an employee's position with regard to employment.

empty homes agency Government funded agency which works with local authorities, private landlords and registered social landlords to bring empty properties into housing use.

enabling Act or Statute An Act of Parliament which allows a local, or public authority or government department to use its powers of compulsory purchase to acquire land for particular specified purposes.

encroachment Unauthorised intrusion of a building or improvement (such as a wall, fence, etc.) onto another's land. Fences, or other structures, that extend into the property of another owner.

encumbrance A legal right or interest in land that affects a good or clear title and may diminish the land's value. It can take numerous forms, such as ordinances, easement rights, claims, mortgages, liens, charges, a pending legal action, or restrictive covenants. An encumbrance does not legally prevent the transfer of real property. It is up to the buyer to determine whether to purchase with the encumbrance.

endorsement The signature on the back of a bill of exchange or cheque making it payable to the person who signed it.

endowment (1) Giving a dower. (2) Provision for a charity.

endowment trust A portfolio or fund established and held in trust for the maintenance, repair and upkeep of usually heritage property.

energy management The systematic and rational approach to the provision and conservation of the use of energy in a building. Often associated with careful measurement and monitoring to ensure that any changes increase energy efficiency.

enforcement notice A notice served by the local planning authority under the Town and Country Planning Act 1990, as amended, requiring that development which is unauthorised must cease or be removed. If it is not then it becomes a criminal offence. (See Planning and Compensation Act 1991.)

enfranchisement A tenant holding a lease exceeding 21 years at a rent of less than two-thirds of the rateable value may acquire the freehold from the landlord or extend the lease.

engineer A qualified person who would be responsible for producing part of the design information such as structural details (i.e. Structural Engineer), electrical details (i.e. Electrical Engineer) and heating and services details (i.e. Heating Engineer).

English Heritage Government quango with responsibility for historic buildings and monuments. The organisation provides advice on planning and conservation matters and is able to provide grants to facilitate conservation.

English House Condition Survey A five-yearly survey of the condition of the English housing stock carried out by the government.

engrossing Preparation of the final version of a deed in writing or print for execution.

Enterprise Zone Government designated area in which the aim to restore private sector activity is facilitated by the removal of certain tax burdens and the relaxation of statutory controls.

entry Going onto land with a clear intention of asserting a right on it.

envelope (1) The parts of a building or structure which enclose it; for example, roof, walls, windows, doors, which together form its external envelope. (2) In planning terms, often referred to the boundary around a particular piece of land.

environmental envelope The enclosing element of a building separating natural and artificial environments which, as such, must satisfy certain criteria in terms of appearance, strength, stability, durability and fire, thermal and sound resistance.

environmental impact statement An evaluation of all aspects and effects a development will have on the environment of a proposed site.

epitome of title   Listing of documents to establish the root of a title to land.

equal opportunities policy   Policies designed to ensure that all are treated equally and fairly in the provision of services and employment.

equitable   (1) Just and fair. (2) In accordance with the practice and procedure of the courts of equity. (3) In accordance with equitable rules.

equitable easement   An easement other than one created by a deed, prescription or statute; e.g. an easement lasting for life.

equitable estate   An exclusive equitable right to land, but since 1925 referred to as an equitable interest.

equitable interest   A charge, or interest on land, other than a legal estate.

equitable remedies   Those remedies principally evolved by equity, for example specific performance, injunctions, delivery up and rescission.

equitable rights   Those rights originally recognised and enforced only in courts of equity. The rights are good against all persons save the bona fide purchaser of value without notice of a legal estate, compare and contrast legal rights which are good against the whole world.

equity   (1) A body of law designed to remedy disputes that could not be heard under common law; where there is any conflict between the rules of common law and the rules of equity, the latter are to prevail. (2) A colloquial term describing the inherent value of a property which is realised when it is sold after paying off the liabilities.

equity capital   The part of the share capital of a company owned by ordinary shareholders.

equity finance   Sums of money advanced in the form of a loan normally made available to a company for a specific venture, such as the development of land. The contract normally entitles the person to a share of any profit.

equity linked mortgage   A mortgage whereby the interest on the principal, in part or in whole, is calculated, usually yearly by reference to changes in the annual return on the security.

equity, maxims of Statements or aphorisms purporting to state some of the fundamental principles of equity; some examples are equity acts in personam, equity follows the law, equity delights in equality, he who comes into equity must come with clean hands.

escheat Reversion of property to the State due to failure to find persons legally entitled to hold or lack of heirs. The State must try to find heirs.

escrow The deposit of instruments and/or funds into the care of a neutral third party with instructions to carry out the provisions of an agreement or contract once all instruments and/or funds have been deposited.

estate (1) An interest in land. (2) The right to use land for a certain period of time. (3) A section of land normally in one ownership; for example, landed estates, a university estate, etc. (4) The total assets of a person, including real property, at the time of death.

Estate Action Programme Programme introduced by the Department of the Environment to fund the refurbishment and rebuilding of inner city housing estates owned by local authorities and registered social landlords. This has now been incorporated into the single regeneration budget (SRB).

estate agency work Under Section 1 of the Estate Agents Act 1979 defined as 'things done by a person in the course of business pursuant to instructions received from a client who wishes to dispose of, or acquire an interest in land'.

estate agent A person who advises the principal in respect of the sale, purchase, lettings, mortgages, etc. See Estate Agents Act 1979, as amended.

estate at will Possession of property at the discretion of the owner.

estate duty Until 1975 a tax that was levied on the value of property owned by a deceased person. (See inheritance tax.)

estate for years Tenant has rights in real property for a designated number of years.

estate management board Tenant led organisation which carries out management functions within social housing.

estate owner The owner of a legal estate in land.

estate renewal challenge fund Fund for which local authorities who wish to transfer ownership of substandard housing estates to local housing companies bid on an annual basis.

estate tail An estate enduring so long as the original owner has a lineal descendant on his death.

estate terrier An estate management information system used for the efficient day-to-day running of an estate. Usually includes details of estate boundaries, details of titles, lettings, easements, etc.

estate, legal Under Section 1 of the Law of Property Act 1925 the only legal estates capable of subsisting are an estate in fee simple (freehold) and a term of years absolute (leasehold).

estoppel An impediment to an action at law. A rule that prevents a person from denying the truth of a statement she has made or from denying facts alleged, when the statement has been acted upon, usually to her disadvantage by another party.

ethnic monitoring Maintenance of record systems which distinguish between ethnic and racial groups so that the results can be analysed to ensure that the organisation is acting in a fair and non-discriminatory manner.

euro bond Bond issued in euro currency.

euro currency Unit of currency held in a European country other than its country of origin, e.g. euro dollars, euro sterling, euro yen.

European Commission Executive arm of the EU consisting of a president and representative commissioners from each member country. The Commission initiates and influences EU legislation and mediates between member governments.

European currency unit Unit of account with which intra EU settlements can be effected.

European Investment Bank Non-profit making bank set up under the Treaty of Rome in 1958 to finance capital investment projects within the European Economic Community. The bank grants loans to private and public organisations for projects which further the aims of the European Community.

European monetary system Established in 1979 as an exchange rate stabilisation system involving the countries of the European Union. Within the system participating countries committed themselves to the exchange rate mechanism which maintained currency values within agreed limits.

European Union (EU) Created in 1993 from the European Community. The title derives from the 1958 Treaty of Rome. The

69

executive body of the EU is the European Commission which was formed in 1967 with the Council of the European Communities. The European Parliament, which was formed in 1957, exercises democratic control over policy and the European Court of Justice imposes the rule of law on the EU as set out in its various treaties.

evasion A colloquial term often used to mean the illegal act or omission resulting from the production of false information in relation to taxation.

evergreen A type of short-term loan offered by merchant banks in which a fixed sum is borrowed for a rolling period, with an annual option to sustain it for a further period and with variable interest rates.

eviction A legal procedure to remove an occupier of land or tenant for lawful reasons including failure to pay mortgage or rent. The eviction of a residential occupier is a criminal offence unless resulting from court proceedings.

ex gratia Something carried out without legal obligation or the admission of liability.

ex gratia payment A sum of money paid usually by a public authority which has no strict entitlement to compensation.

ex parte Application to court by an interested person who is not a party.

examination of title An inspection of public records and other documents to determine the chain of ownership of a property.

exception reporting The reporting to management of items which fall outside given criteria, e.g. arrears in excess of £1,000.

excess charge A sum levied by the landlord at the end of the year for the amounts due and recoverable legally from the tenant under the terms of the lease applied in relation to the provision of services provided for the tenant during the said period.

exchange control Restriction on the sale and purchase of foreign exchange.

exchange equalisation account Account managed by the Bank of England on behalf of the government which contains the official gold and foreign exchange reserves of the UK and is used to control and stabilise the value of sterling on the foreign exchange market.

exchange of contracts The initial formal and legally enforceable step in the transfer of land or property.

exclusive rent   The rent under a lease which makes the tenant responsible solely for the payment of rates, services and other outgoings.

executed   That which is done or completed, for example executed consideration.

executed trust   One where the trust estate is completely defined in the first instance, no other instrument of conveyance being contemplated.

execution   Methods of enforcement of a judgement in an action. The signing of an instrument in a manner which gives it legally valid form.

execution of wills   No will is valid unless in writing and signed in order to give effect to it by the testator or by some other person in his presence and by his direction and the signature must be made or acknowledged by the testator in the presence of two or more witnesses present at the same time and each witness must attest and subscribe the will in the presence of the testator but not necessarily in the presence of any other witness. See the Wills Act 1837 section 9 as amended by the Administration of Justice Act 1982 section 17.

executive director   A director of a company who is also an employee of that company.

executor   One appointed by a will to administer the testator's property and to carry out the provisions of that will.

executor de son0 tort   One who without any authority intermeddles with the goods of a deceased person as if he has been duly appointed executor. He may be sued by the rightful executor, administrator, creditor or beneficiary. See the Administration of Estates Act 1925 sections 25 and 55.

executor, duties of   Getting in the assets of the deceased, paying funeral expenses, paying legacies, accounting for residual estate.

executor's year   The period of one year from the death of the deceased in which the executor must complete the administration of the assets.

executory trust   One where the party who has benefited is to take through the medium of a future instrument of conveyance to be executed for the purpose.

executrix   Feminine of executor.

exempt lease   A lease exempted from the security of tenure provisions of the Landlord and Tenant Act 1954 following a joint application to the court by the landlord and the tenants.

exemption The mechanism which removes legal obligations or financial liability imposed by common law, or statute.

exhibit A document used in evidence, especially when annexed to an affidavit.

existing use rights A colloquial term used to describe the uses of land and buildings which do not require planning permission. The uses are those contained in the use classes order, the general development order or have existing use certificates attaching to them. (See existing use value, use classes order, general development order.)

existing use value (1) A colloquial term meaning the open market value of land or buildings applied to their current use as opposed to any potential use. (2) The value which may apply to a situation where compensation is payable when the Secretary of State declines to confirm a purchase notice and instead directs that planning permission should be granted for some other form of development as per the Town and Country Planning Act 1990, as amended by the Planning and Compensation Act 1991.

expectant heir One who has a prospect of coming into property on the death of another person.

expert A person having particular specialised knowledge.

express That which is not left to implication, for example an express covenant or promise.

express trust A trust created as a result of the settler's express fair dealing rule, a principle related to the nature of trusteeship whereby a trustee may not buy from his cestui que trust (beneficiary) unless this was intended by the beneficiary and there is no concealment or fraud. The principle applies to any person in a fiduciary position, for example, solicitors and clients.

expropriation Compulsory acquisition of an estate or part of an estate by the nation.

external audit An independent examination of the financial accounts of an organisation.

external valuer Under the Royal Institution of Chartered Surveyors Guidance Notes on the Valuation of Assets, defined as a qualified valuer who is not an internal valuer and where neither he nor any of his partners or co-directors are directors or employees of the company, or of another company within a group of companies, or having a significant financial interest in the company or group, or where neither the company nor the group has a significant financial interest in the valuer's firm or company. The company may include a housing association, development company, etc.

extinguishment   The termination of a right or obligation.

F

facade   The front elevation or external face of a building.

face value   Nominal or part value of an item, e.g. that printed on a bank note, coin or security. This may be more or less than the market value.

factoring   The practice of buying the debts of a manufacturer, assuming the responsibility for debt collection and accepting the credit risks.

fair market value   The hypothetical probable price that could be obtained for a property by informed purchasers.

fair rent   The fixing by a rent officer under various Rent Acts of a 'fair rent' for certain types of residential tenancy.

fee simple   (1) A freehold interest in land. The highest possible degree of ownership of land. The estate allows owners to have unrestricted powers to dispose of property, and which can be left by will or inherited. This type of ownership is the maximum interest a person can have in a piece of land. It entitles the owner to use the property in any manner they see fit. (2) The highest form of tenure allowed under the feudal system allowing the tenant to sell or convey by will or be transferred to an heir in the event that the owner died intestate. In modern law, almost all land is held in fee simple and this is as close as one can get to absolute ownership in common law.

fee simple defeasible   The owner of the property holds a fee simple title contingent upon certain conditions.

fee tail   A form of tenure under the feudal system that could only be transferred to a lineal descendant. If there were no lineal descendants upon the death of the tenant, the land reverted to the lord.

fellow   The most senior member of a professional organisation; for example, Fellow of the Royal Institution of Chartered Surveyors.

feoffee   In feudal society a vassal granted land by his lord.

feoffment In feudal society, a lord's act granting a fief to his man.

feu Perpetual lease at a fixed rent.

feudal system A social structure that existed throughout Middle Ages. Tenants leased the land from the local lord (who themselves held under tenure granted by the Crown) in exchange for loyalty and goods or services, such as military assistance or money.

fi fa Abbreviation of fiera facias, execution of judgement by the seizing and selling of the debtor's goods.

fidelity guarantee An insurance against financial loss resulting from fraud or dishonesty committed by officers or employees of housing associations.

fiduciary estate The estate or interest of a trustee in lands or money as opposed to the beneficial interest or enjoyment held by the beneficiary.

fiduciary involving trust or confidence A fiduciary is a person in a position of trust or responsibility with specific duties to act in the best interest of a client; for example, the relationship of trust that buyers and sellers expect from their agent. The term also applies to legal and business relationships. A fiduciary has the rights and powers that normally belong to another person. A fiduciary should exercise the rights with a high standard of care in protecting or promoting the interests of the beneficiary. Other fiduciary relationships exist as between solicitor and client, and principal and agent.

fiduciary issue Part of the issue of bank notes by the Bank of England which is backed by government securities rather than gold.

fiduciary relationship A relationship based on trust in which one person, such as a trustee, is under an obligation to act solely for the benefit of another person.

final account The agreed statement of the amount of money to be paid at the end of a building contract. Company accounts produced at the end of the financial year.

final certificate A certificate indicating the completion of a building contract, normally issued by an architect or surveyor.

Finance Act Annual UK Act of Parliament which gives effect to the proposals made by the Chancellor of the Exchequer in the Budget speech.

financial accounting Branch of accounting concerned with classification, measurement and recording of the transactions of an organisation and so its finance structure and liquidity.

financial futures Futures contract in currencies or interest rates which are bought and sold on specialised markets. In the UK financial futures and options are traded on the London International Financial Futures and Options Exchange (LIFFE).

financial institution Organisation which collects funds from individual organisations or government agencies and invests these funds or lends them to borrowers.

financial provision orders Court orders attached to the granting of decrees of divorce, nullity or separation. Payments by one spouse to another or the children. Under the Matrimonial and Family Proceedings Act 1984 and Child Support Act 1991, as amended, the court in making the orders for periodic or lump sum payments must give first consideration to the welfare of children under 18 and if possible seek to achieve a clean break.

Financial Reporting Council Set up in 1990 in response to the Dearing Report to oversee and support the work of the Accounting Standards Board, the Financial Reporting Review Panel and the Urgent Issues Task Force.

Financial Reporting Review Panel Subsidiary of the Financial Reporting Council which investigates departures from the accounting requirement of the Companies Act 1985 and is empowered to take legal action to remedy departures.

Financial Services Act 1986 UK Act of Parliament which introduced legislation to regulate the provision of investment advice and the sale of investment products within the UK.

financial statement Annual statement of the company's performance which includes the profit and loss account, balance sheet, statement of recognised gains and losses, the cash flow and supporting notes.

Financial Statistics Monthly publication of the UK Central Statistical Office which gives a full account of financial statistics.

Financial Times Share Indexes Number of share indexes published by the Financial Times which show share prices on the London Stock Exchange. The Financial Times ordinary share index (FT30 share index) shows movement of shares in the 30 leading industrial or commercial shares chosen to be representative of British industry.

The Financial Times actuary's shares indexes reflect movement in different sectors of the market and the FT actuary's all share index reflects the movement of some 800 shares and fixed interest stocks. The most well known is the Financial Times Stock Exchange 100 Share Index (the FT-SE or Footsie) which reflects share price movements of the largest 100 companies.

financial viability returns In a housing association context, the housing corporation requires all housing associations seeking a development allocation to produce financial viability returns. They are intended to demonstrate the extent to which associations have the resources to cover the financial risks involved in undertaking various types of development and taking out long-term loans.

financial year (1) The period of 12 calendar months from 1 April to 31 March. (2) For other purposes, often a period of 12 calendar months stated in the articles or memorandums of agreement of a company.

finder's fee A sum payable to an agent employed to find a property on behalf of a client.

first mortgage A mortgage holding priority over the claims of subsequent lenders against the same property.

fit for habitation Implied statutory covenant attached to certain low rent tenancies.

fittings Items which do not form part of the land and are not included as part of the property on the sale of land unless the seller expressly agrees to leave them behind.

fixed asset (capital asset) Business assets such as land and machinery which are acquired for continued use in early profit. These are written off against profits over their anticipated life.

fixed charge A charge paid by the lender as security for a loan.

fixed costs Costs which are constant irrespective of changes in the levels of production or sales. Fixed costs will include business rates, rent and salaries.

fixed interest security Security which gives a fixed stated interest payment. These include gilt edged securities, bonds, preference shares and debentures.

fixed price contract The situation where the total price is fixed at the beginning of a building contract, subject to variations, or fluctuations in stated circumstances.

fixed rate mortgage A home loan with an interest rate that will remain at a specific rate for the term of the loan.

fixed rent A rent which cannot be moved upwards or downwards during the stated time period in a lease.

fixed term A lease or tenancy for a set period. Date of commencement and termination must be agreed before there can be a legally binding agreement.

fixed trust A trust where the interests of the beneficiaries are fixed from the outset.

fixture Personal property permanently affixed to structures or land, usually in such manner that they cannot be independently moved without damage to themselves or the property.

flat The subdivision of premises in residential occupation usually on the same floor and forming part of the same building, divided horizontally from some other part of it.

flat fee A set fee charged by a broker or other agent instead of a commission.

flexible mortgage A mortgage in which the initial repayments are negotiable, usually maintained at a low level and can be switched to a fixed interest mortgage with specified times during the longer term.

flexi-time System of working in which employees are given a degree of flexibility in the hours worked. Provided a number of core hours are worked flexibility exists as to starting and finishing times.

floatation The launching of a public company through share subscription. After floatation shares may be traded on the stock exchange.

floating charge A charge which a lender, for security for a loan, can spread over the borrower's assets.

floating trust A concept arising in Sprange v Barnard whereby a trust agreed between the settlor and first beneficiary/donee is held to be suspended throughout the lifetime of the donee but that will fall down and attach to the property if and when the first beneficiary/donee attempts to alienate the property, e.g. by leaving it to some other person in a will or by selling it later.

floor area The total superficial area of a building including all floors. The Royal Institution of Chartered Surveyors Code of Measuring Practice distinguishes the following categories. Firstly, gross external area; secondly, gross internal area; thirdly, net internal area.

floor space index A procedure used to assist a local planning authority in the control of the density of new development. It is the ratio of gross floor area to site area, plus half the width of any roads which border the property.

flow chart Chart which represents the flow of material, information or people through a system.

forbearance A course of action a lender may pursue to delay foreclosure or legal action against a delinquent borrower.

force majeure French for an act of God; an inevitable, unpredictable act of nature, not dependent on an act of man, e.g. violent storms, earthquakes or lightning.

forced sale value The Royal Institution of Chartered Surveyors Guidance Notes on the Valuation of Assets define it as 'Open market value, but with the proviso that the vendor has imposed a time limit for completion, which cannot be regarded as a reasonable period in which to negotiate the sale, taking into account the nature of the property and the state of the market.'

foreclosure (1) A proceeding in or out of court to extinguish all rights, title and interest of the owner(s) of a property in order to sell the property and satisfy a lien against it. (2) The legal process reserved by a lender to terminate the borrower's interest in a property after a loan has been defaulted.

foreign exchange market International market in which foreign currencies are traded.

foreshore Part of the sea shore lying between mean high and mean low tide lines. It is normally held that the foreshore is owned by the Crown.

forfeiture (1) The process whereby a landlord exercises his right to retake the physical possession of premises and thus to extinguish a lease following the tenant's failure to remedy a breach of the terms of the lease. (2) Statutory power given to a court to deprive an offender of property used or intended to be used for the purposes of crime.

forfeiture of benefit under will A man shall not slay his benefactor and thereby take the bounty, see Re Crippen (1911) and Re Hall (1914), and the Forfeiture Act 1982.

formal tender A formal bid made by a tenderer.

formalities To be valid, a private express trust must comply with the requisite formalities as set out, for example in section 53 Law of Property Act 1925. Statute of Fraud 1677 section 7 replaced by the Law of Property Act 1925 section 53 in relation to formalities for the creating of trusts of land.

forward finance The forward commitment where an amount of money is made available and is repaid relatively quickly thereafter from the proceeds of a sale.

forward letting/sale An arrangement where agreement is entered into for the taking of a lease or purchase of property in advance of completion of the development.

frail elderly housing In the housing association context, housing for frail elderly people. It includes housing which provides personal care and support for this category of person.

frank tenement Freehold.

frank-fee Freehold land.

fraud on a power The power has been exercised for a purpose or with an intention beyond the scope of or not justified by the instrument creating the power, Vatcher v Paull (1915).

fraudulent conveyance A transfer of land without consideration and with intent to defraud a subsequent purchaser. Such actions are voidable by the purchaser under the Law of Property Act 1925.

free market Market free from government intervention.

free trade International flow of goods and services without interference of laws, tariffs, quotas or other restrictions.

freedom from encumbrance Land and property free from any binding rights of parties other than the owner.

freehold The most complete form of ownership of land. A legal estate in fee simple. It differs from leasehold, which allows possession for a limited time. There are varieties of freehold such as fee simple and fee tail.

freeholder A person who owns freehold property rights (i.e. either land or a building).

friendly societies Organisations registered under the Friendly Societies Act 1977, as amended, to provide by voluntary contributions for the relief of members and their families during times of sickness, old age, etc. They are unincorporated mutual insurance associations which possess mortgage lending powers.

fringe benefits Non-monetary benefits granted to an employee and not subjected to tax.

front loading Financial term referring to a loan in which payments are higher in the earlier years of repayment.

front money Sums of money made available usually as initial short-term finance for the development of land. Often the short-term finance involves the interest being rolled up during the construction period of the development project.

frontage The length of a plot of land, or building, measured along the road to which the plot of land or building fronts.

frontager Owner or occupier of land which abuts a highway, seashore or river.

frozen assets Assets which cannot be used or realised.

frustration Determination of a contract by some intervening event, such as destruction of the subject matter.

full disclosure A person must reveal all known facts that may affect the decision of a buyer or tenant.

full management A situation where all aspects of property management are the responsibility of a managing agent.

full rental value The best possible rental that could be expected on the open market.

full repairing and insuring lease A lease where the lessee is responsible for the full cost of repairing, maintaining and insuring the property.

functional obsolescence Impairment of functional capacity or efficiency. For example, homes without indoor plumbing (while they may contain working outdoor plumbing facilities) are considered functionally obsolete. (See also economic obsolescence.)

funding The situation where a lender advances part, or the whole, of the development costs of land, or where a potential purchaser similarly advances part, or the whole of the full development costs, and on completion acquires the property at a price normally calculated with regard to a formula, as embodied in the agreement.

funeral expenses Reasonable expenses involved in burying a deceased person must be paid out of his estate before any other duty or debt, R v Wade (1818).

future estate An estate to take effect in possession at a future time. The expression is frequently used to apply to contingent remainders and executory interests but is also applicable to vested remainders and reversions.

future interest A property right that does not take immediate effect, for example a grant to be for X's life and then to the first of X's daughters who shall attain the age of 30; the interest of the first of these daughters takes effect in the future.

G

G rules The rules provided by the National Federation of Housing Associations for non-charitable housing associations.

garnishee A person against whom a judgement debt is enforced by ordering him to pay a debt owed to the debtor to the judgement creditor instead.

gazumping When a vendor withdraws from an agreed sale of land before a legally enforceable contract exists with a view to securing a higher price from another purchaser.

gazundering When a buyer withdraws from an agreed sale of land before a legally enforceable contract exists with a view to securing a lower price.

geared rent A situation in which rent is determined as a proportion of, firstly, the rental value of, or the actual rent received from, the subject property, or secondly, the rental value of a similar property.

gearing ratios Ratios which express a company's capital gearing. Those based on the balance sheet usually express indebtedness as a percentage of equity. The more the ratio favours liabilities the greater the gearing.

general damages Unascertained damages, to be assessed by the judge.

general development order A statutory instrument permitting certain classes of development without the necessity of making a formal planning application in accordance with the Town and Country Planning Acts.

general improvement areas An area designated by a local housing authority with a view to improving the houses and amenities in older residential areas.

general legacy A bequest in a will which doesn't specifically identify the thing bequeathed, for example a necklace to 'L'.

general lien A lien such as a tax lien or judgement lien, which attaches to all property of the debtor rather than the lien of, for example, a trust deed, which attaches only to a specific property.

general needs index In housing association terms, a statistical index of relative housing needs used by the Department of the Environment to assist them in making housing investment programme capital allocations to local authorities.

general offer Offer of sale made to the general public.

general power (See appointment, power of.)

general warranty deed A deed which conveys not only all the grantor's interests in and title to the property to the grantee, but also warrants that if the title is defective or has a 'cloud' on it (such as mortgage claims, tax liens, title claims, judgements, or mechanic's liens against it) the grantee may hold the grantor liable.

gentrification The rehabilitation of a residential area which was formerly run-down, resulting in the influx of wealthier, often younger, professional peoples moving into the area.

gift (1) The voluntary transfer of property. (2) A cash gift a buyer receives from a relative or other source. Lenders usually require a 'gift letter' stating that the money will not have to be repaid.

gift over A gift which comes into being when a particular preceding estate is determined.

gift, imperfect A gift which has not been completely constituted, i.e. the correct formalities for transfer have not been complied with.

gift, *inter vivos* A grant or transfer of property between living people. Validity depends on the intention to give and appropriate acts to make the intention effective.

gilt edged security (Gilt) Fixed interest security or stock issued by the government of the UK and quoted on the stock exchange.

giro rent collection Facility for tenants to pay rent through a post office giro account or a bank giro account.

going concern concept Accounting principle which assumes that a business is a going concern and will continue to operate for the foreseeable future.

golden handcuffs Incentives offered to persuade staff to remain within an organisation.

golden hello Incentive offered to induce an individual to take up employment with an organisation.

good faith Honesty and decency which can be a requirement in law.

good leasehold title A good leasehold title is granted under the Land Registration Act 1923, as amended, and is equivalent to an absolute title except that it cannot be guaranteed that the landlord will grant a lease. It usually occurs when the deeds showing the title of the landlord have not been registered.

good marketable title The situation where the title is entirely free of encumbrances.

good title The situation where the title to land is supported by adequate evidence.

grace period A specified amount of time to make a loan payment after its due date without penalty.

grant aided land In housing association terms, a phrase used to define property which cannot be disposed of without the consent of the Housing Corporation under Section 9 of the Housing Associations Act 1985.

grant of representation (1) Probate, see later. (2) Simple administration when the deceased dies wholly intestate, i.e. without a will.

grantee A person conveyed an interest in a piece of land. The buyer, who receives the land.

grantor The person who conveys an interest in a piece of land to another person. The seller, who gives the land.

green wedges The concept similar to a greenbelt, where a planning authority designates areas of woodland, agricultural land, or other recreational land for the purpose of separating urban neighbourhoods.

greenbelt The name given to a town planning concept where an area of rural land surrounding a town is designated, and development is discouraged from being undertaken there. The object is to prevent urban sprawl and to provide recreational opportunities for the residents.

grey land The area of land containing property surrounded by, or adjacent to a clearance area where acquisition is necessary for the satisfactory development of the cleared area. It attaches a special basis of compensation for compulsory acquisition.

gross domestic product (GDP) Value of all goods and services produced by an economy over a specified period.

gross external area The aggregate superficial area of a building, taking each floor into account. The Royal Institution of Chartered Surveyors Code of Measuring Practice includes within this concept: external walls and projections, internal walls and partitions, columns, piers, chimney breasts, stairwells and liftwells, tank and plant rooms, fuel stores, whether or not above the main roof level and open-sided covered areas and enclosed car-parking areas that exclude open balconies, open fire escapes, open vehicle parking areas, terraces, etc. and domestic outside WCs and coalhouses. In calculating the gross external area, party walls are measured to the centre-line while areas with a head-room of less than 1.5 metres, are excluded and quoted separately.

gross income The total income of a household before taxes or expenses are subtracted.

gross interest Interest payable on a loan or deposit before tax is deducted.

gross internal area The measurement of a building in the same manner as gross external area, but excluding external wall thick-nesses.

gross lease A lease which requires the lessor to meet all, or part of, the expenses of the lease property such as taxes, maintenance, utilities and insurance.

gross national product (GNP) The sum of gross domestic product and interest, profits and dividends received from abroad by UK residents.

gross value Defined by the General Rate Act 1967 as 'The rent at which a hereditament might reasonably be expected to let from year to year if the tenant undertook to pay all the usual tenant rates and taxes and the landlord undertook to bear the cost of repairs and insurance and other expenses, if any, necessary to maintain the hereditament in a state to command the rent'.

ground lease A long lease granted at ground rent, i.e. a rental value which disregards the value of any buildings on the land, but reflecting the right to develop the land with buildings.

ground rent The amount of money paid for the use of a piece of property when it is a leasehold estate. Rent paid for vacant land. If the property is improved, ground rent is the portion attributable to the land only.

group Group of inter-related companies usually consisting of one holding company and a number of subsidiaries.

group home A residential dwelling where a group of single people live together as one household.

guarantee An agreement to pay a debt in the event of the debt not being paid.

guarantor A third party to a contractual agreement who guarantees to do certain things by a party to the contractual agreement.

H

half commission man Stock exchange term for person who is not a member of the stock exchange but introduces clients in return for an agreed share of the commission.

half secret trust This is created where on the face of the will is disclosed the existence of a trust but not its terms, for example where property is left to 'A' on the trust I have previously discussed with him.

halfblood The blood relationship between people who have only one nearest common ancestor and not a pair of nearest ancestors.

harassment Under the Protection from Eviction Act 1977, as amended, it is an offence for a landlord of residential property to use or threaten violence to obtain possession. The offence can be committed by interference with the tenant's comfort and enjoyment of his home.

hard currency   Currency which is regarded as safe from loss of value and is so commonly accepted internationally.

hazard insurance   Insurance on a property against damages caused by fire, wind, storms, and similar risks.

head rent   The rent payable by the head lessee to the freeholder.

headlease   The principal lease held directly from the freeholder which may be subject to a series of underleases for part, or the whole, of the property.

heads of claim   The categories, or titles, by which claims may be made. Often associated with claims for compensation.

heads of terms   The principal points of agreement which form the basis of a contractual agreement. In a lease they often include the duration of the letting, the initial rent and the rights and obligations of the various parties.

Health & Safety Commission   Commission set up under the Health & Safety at Work Act 1974 to oversee provision of health, safety and welfare of people at work and to protect the public from risks arising from work activities.

hearsay   Testimony by a witness as to a matter not within his personal knowledge

hectare   The equivalent of 2.471 acres.

heir   Previously a person entitled to inherit freehold land from someone who died intestate, now under the Law of Property Act 1925 section 132 a limitation of property in favour of the heir.

heir apparent   The person who will inherit if he or she outlives his or her ancestor, for example the eldest son.

heir presumptive   One whose right to inherit may be superseded by the birth of an heir with a superior claim.

heirs and assigns   One who might inherit or succeed to an interest in a property under the rules of law applicable when a property owner dies.

heirs of the body   Before 1925 descendants who were entitled to inherit freehold land whose owner had died intestate. The laws of primogenitor apply.

hereditament   (1) Real property which devolved onto an heir on intestacy can be classified as incorporeal, for example rights of property, or corporeal physical objects, such as land or buildings. (2) A unit of land that was separately assessed for rating purposes.

heritage property Property of historical value as defined by the Inland Revenue under Section 30 of the Inheritance Tax Act 1984, as amended. It includes: (1) Pictures, works of art, etc which are of national importance. (2) Land of outstanding scenic, historic or scientific value. (3) Building needing special steps for its preservation because of its outstanding interest. Transfer of all such property are potentially exempt from inheritance tax.

high court The principal court of England and Scotland.

highrise building Generally a building of considerable height in relation to the average height of other buildings in the locality. Generally accepted as a block of flats over five storeys in height.

highway Normally understood to mean a main road, but in law, considered as a strip of land over which the public have a right of passage for purposes specified in relation to the particular set of circumstances.

historical cost The cost of an asset based on its original cost rather than current saleable or replacement value.

historical cost accounting System of accounting based primarily on the original costs incurred. Under this system the monetary amounts contained in published accounts are of historical relevance only and do not represent the real cost in terms of the contemporary value. In times of high inflation this can lead to a misleading picture of the true state of the business.

holding company Company within a group which holds shares in other companies.

holding over Continued occupation by a tenant after determination of his lease. If the landlord accepts rent from the tenant a new tenancy is created.

holograph A document, for example a deed or a will, written entirely in the author's handwriting.

homeless families initiative A Housing Corporation programme which attempts to provide homes specifically for homeless families.

homeless person Under the Housing (Homeless Persons) Act 1977, as amended, a person is considered homeless if he has no living accommodation or he has accommodation which he is unable to

87

secure access to. A local authority has a duty to house an unintentionally homeless person.

homelessness Defined under Section 58 of the Housing Act 1985 as one who has no accommodation in England, Wales or Scotland, or who has accommodation but is unable to occupy it. Local authorities have responsibility for housing people who are homeless and in priority need.

homeowner's policy Policy that expands the insurance for a homeowner. It may include theft, liability, subsidence, etc.

hope value The value of a property above its existing use value, determined by the prospect of an alteration giving a more valuable future use. Often associated with the grant of planning permission for a more beneficial use.

hostel A building providing residential accommodation and either board or facilities for the preparation of food.

hostel deficit grant In the housing association context, a grant payable towards deficits arising from the running costs of approved hostel projects, replaced by the Transitional Special Needs Management Allowance.

hotchpot The rule that the estate of someone who dies intestate which is to be divided equally among the beneficiaries takes account of benefits received by any of the beneficiaries before the intestate's death. See also the Administration of Estates Act 1925 section 47.

House of Lords The highest court of England and Scotland.

Housing Act 1964 Act which established the Housing Corporation and co-ownership housing.

Housing Act 1969 Act which increased subsidy for housing associations and introduced general improvement areas.

Housing Act 1972 Act which introduced fair rents for housing associations and provided subsidy for housing association new build schemes.

Housing Act 1974 Act which extended the powers of the Housing Corporation, introduced Social Housing Grant, Deficit Grants and Housing Action Areas.

Housing Act 1980 Introduced the Tenants' Charter and statutory right to buy (i.e. the right given to some tenants to purchase the dwelling in which they are living at a discount after a minimum period of residence) to tenants of non-charitable registered housing associations and local authorities. This Act also introduced the statutory form of accounts, which is the form in which statutory housing association accounts must be published.

Housing Act 1985 The main Act which consolidated housing legislation since 1957, when the last previous consolidation took place.

Housing Act 1988 This Act introduced two types of tenancy into the private residential sector: assured tenancy and assured shorthold tenancy. The former offers wide-ranging security for the tenant, the latter offers no security of tenure. The landlord is guaranteed possession after certain specified periods of time.

Housing Act 1996 Act which made provision for new types of social landlords to register with the Housing Corporation, allowed introductory tenancies within local authorities and significantly altered the rights of homeless persons.

housing action area An area of residential uses declared by the local housing authority under the Housing Act 1985, as amended. It is designated because the condition of the properties are considered unsatisfactory with regard to the physical state of the properties and the social condition of the persons living in the area. It allows authorities to acquire land and repair and improve housing accommodation and the environment, with grants supplied by central government.

housing action trust An organisation established under the Housing Act 1988, as amended, which removes housing from local authority control, handing it over to a statutory trust charged with improving the stock.

Housing and Building Control Act 1984 This Act introduced transferable discounts for tenants of charities and two new rights for secure tenants (i.e. local authority tenants and housing association tenants whose tenancies commenced before 15.1.89); the right to repair and the right to exchange.

housing association Under Section 1(1) of the Housing Associations Act 1985 and Section 5(1) of the Housing Act 1985 defined as 'a society, body of trustees or company, which is established for the purpose of, or amongst whose objects or powers are, including those who are providing, constructing, improving or managing, or facilitating, or encouraging the construction or improvement of, housing accommodation and which does not trade for profit or whose constitution or rule prohibits the issue of capital with interest or dividends exceeding such rate as may be prescribed by the treasury whether with, or without, differentiation as between share and loan capital'.

housing association grant   The main form of financial subsidy for housing association work paid for by the Housing Corporation.

housing association tenancy   A tenancy where the landlord is a housing association, the Housing Corporation or housing trust.

Housing Associations Act 1985   One of three consolidating Acts passed in this year. This Act contains all the legislation passed to date which relates specifically to housing associations.

housing authority   Under Section 4 of the Housing Act 1985, defined as 'a local housing authority, a New Town Corporation, or the Development Board for Rural Wales'.

housing co-operative   Under Section 27(2) of the Housing Act 1985 defined as 'a society, company, or a body of trustees for the time being approved by the Secretary of State for the purposes of this section', which empowers a local housing authority to enter into agreements for co-operatives to undertake any of the housing authority's duties concerning the provision of housing accommodation on land owned by the authority.

Housing Corporation   An organisation established under the Housing Act 1964 which is responsible for promoting non-profit making housing associations, with a view to providing homes for people in need. It maintains the register of housing associations and makes public funds available for such associations to develop appropriate housing developments.

Housing Homeless Persons Act 1977   Legislation made local authorities responsible for providing accommodation for those persons accepted as being homeless. Amended by Part III of Housing Act 1985.

housing investment programme   The annual programme of capital expenditure on housing for each local authority, agreed by the then Department of the Environment.

housing manager   A person who engages in the profession of the management and maintenance of properties.

housing needs indicator   A measure of the relative requirement within each local authority area for new provision of housing by housing associations for people in need. It is used by the Housing

Corporation to decide the allocation of capital resources between its regions and gives guidance for expenditure within regions.

Housing Organisations Mobility and Exchange Scheme In housing association context, a national organisation funded by the then Department of the Environment and formed in 1990 with responsibilities to streamlining exchanges among people living in social housing.

housing plus Term used to describe activities beyond basic housing which give extra support to individuals or to the community through provision of employment opportunities, community facilities, etc.

housing repairs account The local authorities account for repairs purposes.

housing revenue account The landlord's revenue account for local authorities.

housing revenue account subsidy The then Department of the Environment's subsidy paid to local housing authorities according to an annually determined formula.

housing society An organisation registered under the Industrial and Provident Society's Act of 1893 which does not trade for profit and whose primary function is for the purpose of improving, constructing or managing houses.

housing trust Under Section 2 of the Housing Associations Act 1985, and Section 6 of the Housing Act 1985, defined as 'a corporation, or a body of persons which (a) is required by the terms of its constitution to use the whole of its funds, including any surplus which may arise from its operation, for the purpose of providing housing accommodation or, (b) is required by the terms of its constitution instrument, to devote the whole, or substantially the whole, of its funds for charitable purposes and in fact, uses the whole, or substantially the whole of its funds, for the purpose of providing housing accommodation'.

human resource accounting Attempt to attach a monetary value to the human resources of an organisation and show them on the balance sheet.

human resource management Management of people within an organisation in order to achieve individual behaviour and performance which will enhance organisational effectiveness.

hyper inflation Extremely high level of inflation.

I

illegal trust A trust which offends against morality, public policy or statute.

illusory trust A transfer by a debtor to trustees upon trust for the creditors. In some circumstances the debtor can revoke it.

immediate landlord Normally held to be the freeholder, head lessee and/or occupying tenant.

imperfect gift (See gift, imperfect.)

imperfect obligations Moral duties which are not enforceable at law, for example charity or gratitude.

imperfect rights (See perfect rights.)

imperfect trust An executory trust.

impersonal account Term used within double-entry book keeping to describe accounts which do not bear the name of a person.

implied Suggested or understood by implication or deduction from the surrounding circumstances; thus an implied trust has been defined as 'a trust which is founded in an unexpressed but presumable intention'. Example: 'B' purchases property in the name of 'C'. There is a presumption in equity that 'B' intended 'C' to hold that property in trust for him. Contrast this with a constructive trust, which is one construed by a court of equity without reference to the presumed or actual intention of any party.

implied covenant A covenant which is assumed by the law to exist in a lease even though it may not be so expressly stated.

implied trust (See implied.)

imports Goods or services purchased from other countries.

imprest account Petty cash system in which cashier is given a fixed float or imprest for sundry expenses. At the end of a fixed period or when the money is spent the cashier provides appropriate vouchers or receipts for the amount spent and is reimbursed so that the float or imprest is restored.

improved ground rent A rent which does not exceed the full rental value of land, but is greater than the previously paid ground rent under a lease, since the grant of which land values have increased.

improvement for sale   In the housing association context, a scheme where housing associations can buy and modernise older property for outright sale containing a limited payment of Housing Association Grant.

improvement grants   Grants which are means tested and paid by local authorities to owner occupiers to assist them in improving their property to an agreed standard.

improvement line   The delineation on a plan alongside a street, and some distance from it, where buildings would not be permitted to be erected or extended. The line is normally designated for the purpose of allowing road widening.

improvement notice   Under the Housing Act 1985, a notice issued by the local housing authority requiring a dwelling which has fallen below set standards, to be brought up to an acceptable standard. The procedure can only be implemented if the residential property is contained within a general improvement area, or housing action area, or has been built or converted before 3 October 1961, and representations have been made by the occupying tenant and the local housing authority in an agreement that it is below the required standard.

improvements   Valuable additions to property that raise the value of the property.

in loco parentis   In the place of a parent.

in personam   The nature of a claim or action brought against a specific person or a right that affects a particular person. (See also *in rem*.)

in rem   Literally 'against the thing.' The words describe a right such as the ownership of property as well as an action against a piece of property, as opposed to a person or persons. (See also *in personam*.)

in the market   Property which is presently for sale or to let; also describes a person looking to purchase or rent a property.

inalienability, rule against   Known also as the rule against perpetual trusts. The general principle is that property must not be rendered inalienable.

inalienable   Not transferable.

incapacity   A lack of the necessary legal confidence to, for example, make a valid contract or trust.

incentive fee   (See contingency fee.)

inclusive rent A rent payable where the landlord has an obligation to pay the rates. It may also include the situation when a landlord is responsible for the provision of certain services.

income and expenditure account Account of income received and amounts expended by an organisation or an individual over a given period.

income support Benefit designed to supplement low income. It is means tested and does not include housing benefit. A person is entitled if their income is below a specified amount and the person is actively seeking employment.

income tax A levy on income or profits administered under the Income and Corporation Taxes Act 1988, as amended. The tax is implemented each year under the annual Finance Act. The tax is related to different types of sources of income such as: rents, company dividends, etc. The tax year of assessment runs from 6 April to 5 April. The tax is managed by the Commissioners of Inland Revenue, Inspectors of Taxes being their subordinate local officers. A person who is aggrieved by an assessment made by the Inspector may appeal to the local Commissioner.

income tax allowances Allowances which may be deducted from an individual's gross income before calculating the income tax liability. Examples of allowances include personal allowance, married couples' allowance and widow's bereavement allowance.

income tax code Code issued by the Inland Revenue which takes account of the income tax allowances of an individual. The code is used by the employer to calculate tax payable.

incompletely constituted trust A trust which requires some other action by the settler before it is perfectly created.

Incorporated Association of Architects and Surveyors (IAAS) A body established in 1925 with a view to registering the interests of its members, each of whom was practising as an architect and/or surveyor. Its members are required to abide by a code of practice and be qualified by professional examinations.

Incorporated Society of Valuers and Auctioneers (ISVA) An organisation established in 1967 by the amalgamation of the Valuers' Institution and the Incorporated Society of Auctioneers and Landed Property Agents. Its members have to abide by a code of professional practice and election is by way of professional examinations.

incorporation The merging together into a single whole. Referring mainly to the creation of a single legal personality.

incorporation by reference Where a testator in a duly executed will refers to an existing document which remains unattested then that document becomes part of his will; in other words it is incorporated into it.

incorporeal Rights and interests which are intangible are said to be incorporeal, such as debts or shares in a company.

incremental expenditure Expenditure incurred in pursuing a project which is additional to that involved in not pursuing the project.

incurable defect A defect in a property that cannot be fixed, such as an adjacent hazardous waste site, or would cost too much to repair relative to the value of the property.

indefeasible That which cannot be made void.

indemnity An agreement made by one person to make good any loss suffered by another.

indemnity insurance Protection resulting from an insurance policy (contract) whereby on the occurrence of the insured risk, for example, flood, the insurer will make sufficient payment to meet the financial losses incurred.

indemnity period The term by which an insurance policy will entitle the holder to receive compensation for loss, injury or damage, arising from the risk defined in the policy document.

independent expert A professional person with appropriate specialist knowledge appointed to resolve a difference between two parties. He examines the evidence presented by the two parties, and can use his own expert knowledge to resolve the dispute. His decision is normally binding on both parties, unless one party can prove negligence. Sometimes referred to as an arbitrator.

Independent Housing Ombudsman New name for housing association tenants' ombudsman as from 1 April 1997.

independent living fund Benefits Agency fund set up to enable disabled people to pay for support and carers in their own homes.

independent surveyor A surveyor who is impartial and appointed with a view to resolving a dispute. The surveyor may be appointed as an arbitrator, or simply to express an independent opinion, either expressly or implicitly, depending on the terms of his particular appointment and the circumstances of the dispute.

independent valuer A surveyor engaged to undertake an impartial valuation. The appointment may be to adjudicate between two parties, or to settle a difference in valuation between two existing valuations.

indexation Practice of linking economic variables such as wages, taxes and pensions to rises in the general price levels.

index-linked mortgage Low start mortgage in which the interest rate and loan term are fixed at the start. Part of the principal is repaid each year and the remainder is index-linked.

indicator rents Guide lines on rents issued to housing associations by the National Federation of Housing Associations. They are revised annually.

indirect construction costs Costs incurred in building works which are beyond the normal ones of labour, machinery and materials. Normally taken to include the cost of financing, administration, insurance, taxes and loss.

indirect costs Costs which cannot be directly attributed to units of production or other organisational activities. These may include property costs, administration costs and costs associated with selling.

indorsement Something written on the back of a document.

Industrial and Provident Societies Organisations controlled under the auspices of the Registrar of Friendly Societies which must accord with the Industrial and Provident Societies Acts in organising their affairs. The majority of housing associations are industrial and provident societies.

Industrial Housing Associations Housing associations created particularly to provide residential accommodation for key workers of the workforce of a particular industry and company.

inflationary gap Gap between the total spending within an economy and the amount of spending which would be needed to maintain full employment.

informal tender A tender not containing the characteristics appertaining to a formal tender.

information technology The use of computers to process and distribute information.

infrastructure The services which are required for the development and use of land, normally taken to include roads, bridges, railways, gas, water, sewerage, electricity and telephone installation.

inherent defect A defect contained within the structure of a building which was unintentionally built in during the design or construction, or both.

inheritance (1) The process by which property passes from a deceased person to the beneficiary named in his or her will or through the provisions of the rules of intestacy. (2) Property that a beneficiary receives from the estate of a dead person.

inheritance tax Formerly termed capital transfer tax and being a levy charged on transfers made during the lifetime of the taxpayer as well as on his or her death. Lifetime gifts may be exempt if given seven years before the death of the grantor. If within three to seven years, tax, if payable, is on a sliding scale.

initial return/yield The net income initially received at the date of purchase and usually expressed as a percentage of the purchase price.

injunction A court order that prohibits a person from doing something is a prohibitory injunction, or requires a person to do something is a mandatory injunction. Often an injunction is needed urgently when an interim injunction is granted.

injunction, Mareva Now known as a freezing injunction. (See Mareva injunction.)

injurious affection Reduction in the value of an interest in land, resulting from the implementation of statutory powers.

Inns of Court The four societies of Inner and Middle Temple, Lincoln's Inn and Gray's Inn. Every barrister must be a member of the Inns.

insider dealing Dealing in stocks and shares by an individual who, through access to special knowledge, might influence the value of those stocks and shares.

insolvency Inability on the part of a person or an organisation to pay debts when they fall due.

insolvent The situation where a person is unable to pay their debts when they become due.

inspection In terms of building society valuations, defined by the Royal Institution of Chartered Surveyors, as 'a visit to, and examination of a property, for the purpose of obtaining information prior to expressing a professional opinion as to its value, state of repair, or any other aspects'. The extent of the inspection will depend upon its purpose.

Institute of Housing Former name of the Chartered Institute of Housing.

institutional investors Normally held to include pension funds, insurance companies, banks, unit trusts, etc.

instrument A written legal document such as a deed or a will which has been executed formally.

insurance The situation whereby a person (the insurer) indemnifies the other person (the insured) against financial losses resulting from a loss caused by way of damage or injury suffered for a situation specified in the insurance policy/contract. Owners and buyers can purchase various types of insurance: e.g. hazard, private mortgage and earthquake. The policies guarantee compensation for specific losses.

insurance contract The contractual agreement between the insured and insurer stating the precise terms and conditions of premiums payable and the events against which loss or damage will be compensated.

insurance value The amount of money estimated that would normally indemnify the owner and/or occupier of a building if it is destroyed or damaged. It is normally held to be the cost of reinstatement including professional fees and also possibly some form of indexation.

insured A party which is protected under an insurance policy.

insurer The party providing protection under an insurance policy.

intangible asset An asset that cannot be seen or touched, e.g. goodwill, patents and copyrights.

intangible property (See tangible property.)

intention The term that describes the mental state of a person who aims to cause something to happen.

intention of testator The interpretation that a court seeks to establish of the intentions of a testator in his or her will.

inter vivos Between parties who are alive.

inter vivos **gift** (See gift, *inter vivos*.)

interest (1) A right or title to or a stake in any real or personal property. (2) The return on capital invested. (3) Concern in the outcome of an event.

interest in expectancy A reversionary interest. (See reversion.)

interest in possession An interest which confers a right of present enjoyment on a person.

interest in property The right of ownership, or some right existing in land owned by another party.

interest on unpaid compensation The interest that an acquiring authority must pay under compulsory purchase powers on the out-standing compensation from the date of possession until completion of the conveyance.

interest rate The sum, expressed as a percentage, charged for a loan.

interest-only loan This pays only the interest that accrues on the loan balance each month. Because each payment goes toward interest, the outstanding balance of the loan does not decline with each payment.

interim Provisional until further direction; for example, an interim order.

interim certificate A certificate issued by an architect or surveyor certifying that work carried out under the terms of the contract during a specified period has been completed and requesting payment from the employer to the contractor.

interim control System employed by an organisation to ensure that accuracy is maximised and opportunities for fraud minimised.

interlineation Writing between or on the lines of a document; for example, a deed or a will.

interlocutory A matter dealt with before the trial of an action, such as an interlocutory injunction.

internal rate of return The rate of return at which the cost of a capital investment project and its future cash flows balance.

internal repairing lease The situation where a lease requires that all, or some, of internal repairs, are the responsibility of the tenant. The extent of external repairs to be undertaken by the landlord will be specified in the terms of the lease.

internal valuer A professional valuer employed within a particular organisation to value the assets of the organisation.

international accounting standards Accounting standards issued by the international accounting standards committee.

International Monetary Fund Agency of the United Nations established in 1946 with the objective of maintaining and stabilising rates of exchange, the promotion of international monetary co-operation and the expansion of international trade.

international will A will made in accordance with the annex to the convention providing a uniform law on the form of an international will 1973. See Administration of Justice Act 1982 section 27 subsection 3.

Internet International network of computers connected by modems, lines, telephone cables and satellite links.

Internet service provider A company which sells access to the Internet.

interpretation The determining of the meaning of written documents such as wills or deeds.

interpretation of wills The process of determining the true meaning of a will and the intention of the testator.

interrogatories Questions formally put in writing by one party to another, before the trial of a civil action.

intestacy The situation when a person dies without having made a will. If there is no will a state of total intestacy exists, if a will disposes of only part of the deceased's estate it is known as partial intestacy. There are set rules found in the Administration of Estates Act 1925 governing the disposal of an intestate estate.

intimidation Violence or threats of such, the aim being to compel a person to abstain or to do something which he has a legal right to do or abstain from doing.

intra vires Within the powers.

introductory tenancy Tenancy introduced by the Housing Act 1996 under the terms of which a local authority may evict a tenant within the first year of the tenancy without recourse to court action.

investment The application of a capital sum to purchase an asset with a view, hopefully, of increasing the value of the asset upon subsequent disposal. In relation to trustee investment, see Re Wragg (1919) which defined investment as to apply money in the purchase of some property from which interest or profit is expected and which property is purchased in order to be held for the sake of the income, which it will yield. The Trustee Act 2000 has given trustees statutory rights to invest as an absolute owner.

investment appraisal   Use of accounting and mathematical techniques to establish the likely return of a particular investment project.

investment company   Under Section 103 of the Taxes Act 1988, is defined as 'any company whose business consists wholly, or mainly, in the making of investments and the principal part of whose income is derived therefrom'.

investment file   In the housing association context, it comprises the association's monitoring profile, the association scheme of work profile and the association investment profile and is prepared and used by the Housing Corporation to establish whether a particular housing association is eligible for further capital funding.

investment method   A method of valuation of an interest in land determined by the capitalisation of estimated, or actual, net rental income.

investment property   A property purchased with a view of retaining it and enjoying the total return at some future date, i.e. capital appreciation and/or income over the life of the interest acquired.

investment yield   The return expressed as an annual percentage which is considered to be appropriate for a particular valuation or investment. It is an estimate of the investor's opinion about the prospects and risks attached to the investment. The better the prospects and lower the risks, the lower the expected yield and greater the capital value.

investment, trustees powers of   A trustee may not lawfully invest trust funds upon securities other than those authorised by the trust settlement or statute. See Trustee Act 2000, which permits a trustee to invest as an absolute owner.

investor in land   A person acquiring land with the intention of holding it for a return in the form of rent and/or longer term capital appreciation.

invoice   Document stating amount of money due for goods or services supplied.

issue (1) The children of a man and his wife. (2) Descendants generally. (3) The matter of a court dispute. (4) A point in question. (5) The putting out of banknotes and other paper money for public circulation.

issue price The price at which a new share issue is sold to the public.

issue, dying without (See dying without issue.)

issued share capital Face value of the shares of an entity that have been issued regardless of what price they realise when sold or what price they are currently attracting.

J

JCT contract A standard form of building contract published by the Joint Contracts Tribunal. It is the standard form of contract used in the construction industry.

job analysis Detailed study of a particular job and its relation to other jobs within an organisation. The analysis should lead to the best method of carrying out the job and the qualities needed by the job holder.

job description Written document stating the qualifications and experience required together with an outline of the key tasks and responsibilities attached to a particular post within an organisation.

job evaluation Assessment of an individual job based on the skills, experience and qualifications required to carry out the job and the responsibilities carried by the job holder. This evaluation will be used to determine an appropriate level of remuneration for the job and the differentials between that and other jobs within the same organisation or industry.

job sharing Division of the work of one full-time employee between two or more part-time employees.

joint account In banking an account held in the name of two or more people.

joint agent Two or more agents appointed by the principal to act on his behalf, for the sale or lettings of buildings; two or more estate agents charged by the owner to secure the sale or letting.

joint and several liability Where two or more persons enter into an obligation by which each party is liable severally (individually) as well as jointly with the others.

Joint Contracts Tribunal (JCT) A forum of people representing contractors, surveyors, architects, etc. in order to publish standard forms of building contract.

joint finance In the social housing context finance provided by a district health authority and a social services department of a local authority.

joint funding In the housing association context, an arrangement in which revenue costs in excess of those covered by the management and maintenance of houses are met by some of the statutory charitable sources, before approval for Housing Association Grant is given.

joint heir co-heir. (See heir.)

joint liability The responsibility of two or more people to fulfil the terms of a home loan or debt.

joint mortgage A mortgage taken out by two or more mortgagors.

joint obligation The situation where two or more parties enter into an obligation so that in the event of litigation, all must sue or be sued together.

joint sole agent The situation where two or more estate agents are instructed to sell or let property. It is normal custom for the agents to share a commission fee irrespective of whether the property is sold by the other agent.

joint stock company Company in which a number of traders pool their stock and trade on the basis of joint stock.

joint tenancy A tenancy entered into by two or more parties. Where two or more persons own a tenancy, no one person has a separate share. Four important concepts exist in a joint tenancy: (1) Possession – each must be entitled to possession of the whole property. (2) Title – each must acquire by virtue of the same instrument. (3) Time – each co-owner must acquire the right at the same time. (4) Interest – each must acquire the same interest.

joint tenants Ownership by two or more people that gives equal shares of a piece of property. Rights pass to the surviving owner or owners. Upon death of a joint tenant, the interest passes to the surviving joint tenants rather than to the heirs of the deceased.

joint valuers The situation where two or more valuers are appointed to provide a valuation. (See Royal Institution of Chartered Surveyors Guidance Notes on the Valuation of Assets.)

joint venture The situation where two or more parties combine to carry on a single business for profit.

joint will One document in which two or more people incorporate their testamentary desires. However, it takes effect as separate wills of the persons who have made it.

jointure (1) A joint interest limited to husband and wife. (2) An estate settled on a wife in place of dower.

journal Book keeping term for a book of prime entry where details of transactions are first entered.

judge A public officer appointed to adjudicate on causes in a court of justice.

judgement The decision of the court following legal proceedings.

judgement creditor A party whom judgement is made in favour of, for a sum of money.

judgement debtor A party whom judgement is made against for a sum of money. Such a party may have the judgement enforced by a variety of means if payment is not made, e.g. bankruptcy proceedings, charging order on property, etc.

Judicature Acts (See Supreme Court.)

judicial factor In Scotland a trustee appointed by the court to administer the estate of someone who is incapable of doing it.

judicial foreclosure A procedure to handle foreclosure proceedings as civil matters.

judicial separation A separation of man and wife by the High Court or the County Court which has the effect of making the wife a single woman for all legal purposes except that she cannot marry again. See Matrimonial Causes Act 1973.

judicial trustee A trustee appointed by the court and controlled by the court. See the Judicial Trustees Act 1896.

junk bonds Bonds which carry a relatively high probability of default and therefore offer a high rate of interest in compensation for the risk involved.

jura in personam Rights in personam.

jura in rem Rights *in rem*.

justices of the peace Magistrates are persons appointed by the Crown to act in a quasi-judicial capacity dealing with minor cases.

K

killing, acquisition of property by A beneficiary who by some criminal act kills the testator or next of kin who kills an intestate. These people will not be allowed to benefit from their crimes knowingly with knowledge of the facts in question.

King's Bench A division of the High Court, called Queen's Bench when the sovereign is female.

knowledge Acquaintance with or awareness of truth or fact.

L

labour costs Expenditure on wages to those concerned with the production of a product or service.

labour turnover rate Ratio of the number of employees leaving an organisation or industry within a stated period to the average number of employees working in that organisation or industry within the same period.

laches An unreasonable and unnecessary delay in claiming an equitable right, which denies the plaintiff the assistance of the court.

laissez-faire economy Economy in which market forces are largely unregulated and government intervention is kept to a minimum.

land In a legal sense, the physical ground or soil regarded as the subject of ownership and everything annexed to it, e.g. trees, buildings, fences, etc. and everything in or on it, such as minerals and running water.

land assembly The process of acquiring individual pieces of land to form a larger single unit with a view to its subsequent development or redevelopment.

land availability study An investigation undertaken to establish land which is available for a particular purpose. It is normally undertaken by surveyors, developers, local planning authorities, etc.

land bank The stock of land held by a developer with a view to future development, or redevelopment.

land charges Land charges are administered under the Land Charges Act 1972, as amended. They are kept in a register held by the Land Charges Registry.

Land Compensation Act 1973 This legislation allows that compensation can be obtained if intolerable noise cannot be reduced, and it reduces property values.

land contract Instalment plan for buying a house. It is used as an alternative to obtaining a loan from a traditional source.

land registration The compulsory registration of land on first conveyance of the freehold or grant of a lease greater than 21 years. The record is in three parts: (1) The property register describing the land. (2) The proprietorship register which states the title absolute and any restrictions appertaining to the sale of land. (3) The charges register which indicates the mortgages, and restrictive covenants. The registered proprietor is given a land certificate.

land survey (See survey.)

land surveyor (1) A professional person who normally works for the Ordnance Survey of Great Britain and is responsible for producing ordnance survey maps. (2) An independent surveyor who produces topographical plans of an area of land.

land, compulsory purchase of Land acquired by compulsion under an enabling statute for a stated public purpose and follows standard procedures contained in the Acquisition of Land Act 1981, as amended.

landlord The owner of the freehold or superior leasehold title who grants a lease to the tenant.

Landlord and tenant The contractual relationship between the landlord and tenant. Such formal and informal agreements are controlled by the various Landlord and Tenant Acts.

Landlord and Tenant Act 1954 (as amended) Part 11 of this Act allows tenants to remain in occupation at the end of the tenancy.

Landlord and Tenant Act 1985 This Act codifies the general law on landlords and tenants which is not particular to local authority or housing association dwellings, e.g. repairing obligations, service charges and rent books.

Landlord and Tenant Act 1987 This Act gives tenants of flats with long leases at low rents a collective right to take over any reversion.

Landlord and Tenant (Covenants) Act 1995 Applies to leases granted on or after 1 January 1996. The original landlord (who granted the lease) and original tenant will be bound by all covenants in the lease while the tenant remains a tenant, but when the tenant lawfully assigns the lease he is automatically released from future liability under the lease covenants unless he has agreed to enter an 'authorised guarantee agreement' with the landlord.

Lands Tribunal A body established by the Lands Tribunal Act 1949, its function being to determine questions relating to compensation for compulsory purchase and the discharge of restrictive covenants. It deals with appeals from local valuation courts.

lapse of gift The failure of a bequest because the beneficiary has predeceased the testator. In this event, the gift becomes part of the testator's residuary estate and subject to certain conditions goes to the residual beneficiaries. The doctrine of lapse does not apply to beneficiaries under a secret trust, nor to the devisee of an entailed interest if he or she leaves descendants capable of inheriting who are living at the time of the testator's death, nor to a child devisee or legatee if the child leaves issue at the time of the testator's death. See the Wills Act 1837 as amended by the Administration of Justice Act 1982.

Large Scale Voluntary Transfer Association In the housing association context, an association sponsored by a local authority to take up local authority housing under a Large Scale Voluntary Transfer scheme.

late charge A penalty for failure to pay an instalment on time.

latent damage Damage which exists in a building or structure which is not known to the purchaser of the property at the time of purchase, but subsequently becomes known to the purchaser. It is generally held not to be identifiable by the normal process of inspection.

latent defect A concealed or hidden inherent defect in the design or construction of a building which could not normally be identified by the usual process of inspection.

latent value The potential value accruing to a property on the occurrence of some future event. For example, the granting of planning permission for a greater beneficial use.

law reports Authenticated reports of decided cases in superior courts.

Law Society The governing body of English solicitors.

layout The internal or external grouping or arrangement of a building or buildings.

lease Agreement for the exclusive possession of property for a term of years in return for a periodic rent. It must specify the term of years. The person granting the lease is called the lessor and the person who rents it the lessee. If the lessee grants the lease it is called a sublease. All leases for a period exceeding three years must be executed by way of deed. See Law of Property Act 1925, as amended.

lease with option to purchase A lease under which the lessee has the right to purchase the property. The option may run for the length of the lease or only for a portion of the lease period.

Leasehold Reform Act 1967 This Act gives tenants with long leases of houses (but not flats) the right to acquire the freehold or to take a 50 year extension of their lease, provided they have occupied the house as their only or main residence for the previous three years.

Leasehold Reform, Housing and Urban Development Act 1993 This Act allows tenants with long leases of flats at low rents to have a collective right to compel the landlord to sell the freehold to them.

leaseholds Land held under a lease and transferable by assignment.

ledger In book keeping the ultimate record book showing all transactions entered into by the business using double-entry book keeping. The ledger is usually split into several books; the nominal ledger containing impersonal accounts, the sales ledger containing customer accounts and the purchase ledger containing supplier accounts.

legacy A gift of personal property under the terms of a will. A legacy may be specific, demonstrative or general. (1) A general legacy is one payable out of the general assets of the testator, for example a sum of money or stocks and shares. (2) A specific legacy is the bequest of a specific part of the testator's personal estate, for example a particular piece of furniture or jewellery. (3) A demonstrative legacy is a gift by will of a certain sum directed to be paid out of a specific fund. (4) A residuary legacy refers to all the testator's property after the repayment of any outstanding debts and the disposal of those types of legacy listed above. (See also devise.)

legacy, cumulative A legacy additional to one previously given to the same legatee in the same or subsequent instrument.

legacy, substitution of A gift of personalty by a testator made in lieu of a previous gift, where he or she indicates that he or she does not wish the legatee to take both gifts.

legal description An expanded and unique description of a property that is used on legal documents, such as deeds and deeds of trust. Recorded documents generally require a legal description.

legal rights Rights *in rem*, that are available against the world at large as compared with equitable rights or rights *in personam*.

legal tender Money which if offered must be accepted in discharge of a debt.

legatee A person who is given a legacy.

legatim In Scotland the part of a deceased person's moveable estate by law may be given to the child or children of the deceased, one-third if the spouse survives, one-half otherwise.

legitimacy The legal status of a person whose parents were married at the time of conception or birth.

legitimation Changing a person's legal status from illegitimacy to legitimacy. This can be affected by the marriage of the parents provided the father is resident in England or Wales at the time of marriage.

lender A general term encompassing all mortgages, and beneficiaries under deeds of trust.

lender of last resort The central bank of any economy with responsibility for controlling its banking system. In the UK the Bank of England performs this role.

lender's instructions A document that lenders prepare for the closing agent that outlines the requirements for loan closing.

lessee One who possesses the right to use or occupy a property under lease agreement.

lessor A party who grants a lease.

letter of comfort Letter provided by the third party to one particular party in a contract indicating that the third party is likely to provide funds or a service to assist in whole, or in part, on the successful completion of the contract. Often associated with a letter in principle indicating that funding for a building project will be forthcoming from a bank, building society, or other institutional lender.

letter of intent A formal method of stating that a prospective developer, buyer or lessee is interested in property.

letters of administration In cases of intestacy authority granted to a person by the court to administer the deceased's estate. Letters of administration are also issued if the deceased person did not appoint executors.

letting value Rental value.

liability A debt.

licence An authority to occupy land where the occupant has no interest in the land, in contrast to a lease. The existence of exclusive possession is decisive in determination of whether a licence or lease exists. Licences do not have statutory protection of security of tenure.

licensee A party to whom a licence is granted.

lien The common law right to hold property of another party as security for fulfilment of an obligation, usually associated with payment of a debt.

life beneficiary A person receiving payments from a trust for her lifetime.

life estate An estate for the life of the tenant or by operation of law (See estate.)

life interest A person's interest in property which lasts only for the lifetime of the grantee or the lifetime of another named party.

life or lives in being Used for the purpose of the rule against perpetuities. The common law rule was that the lives in being selected by the donor could be stated expressly or by implication and there is no restriction as to the number of lives selected. The Perpetuities and Accumulations Act 1964 introduced the so-called statutory lives in being.

lifetime gift (See gift, *inter vivos*.)

lifetime homes Design criteria which allow the development of housing to accessible standards, which meets the needs of almost everyone.

light The right acquired under the Prescription Act 1832, as amended, to unobstructed access to light at a person's window. There must be uninterrupted access to the light for 20 years. The Rights of Light Act 1959, as amended, allows the owner of land to prevent the acquisition of a right of light over his land by notice registered in the local land charges register.

light and air easements Easements which entitle an owner, or occupier of property, to the benefit of adequate light and air.

limitation of actions The period within which actions for legal redress have to be commenced.

limited administration Authority to administer the estate of a deceased person for a limited time, for example if a sole executor is a minor, or pending legal proceedings.

limited company A company registered under the Companies Acts, being either public or private, and the financial liability for debts are limited to the value of the shareholding.

limited liability The degree to which debts of a company are the legal responsibility of the shareholders.

limited liability company A company with limited liability.

limited owner Ownership of an interest in property less than freehold.

limited partnership A partnership consisting of one or more general partners who conduct the business, are responsible for losses, and one or more special partners, who contribute capital, and are liable only to the amount contributed.

limited tender The form of tender where the invitation to submit tenders is limited to a specified class, group, or stated tenderers.

line of credit The situation where a borrower is entitled to draw a specified sum of money over a given period of time from a bank on terms agreed at the outset.

lineal consanguinity The relationship between ascendants and descendants, as between father and son, grandmother and granddaughter, etc.

lineal descent Descent in direct genealogical line.

liquid assets Current assets which consist of cash or items which can be quickly converted into cash.

liquidated damages Ascertained or calculated monetary loss claimed in an action. Also a sum provided by a contract as payable in the event of a breach, which is deemed to be a penalty.

liquidation The winding up of a company.

liquidity The extent to which a party can meet financial commitments.

liquidity ratio The ratio of liquid assets to current liabilities. This ratio is regarded as an acid test of an organisation's solvency and is therefore sometimes called the acid test ratio.

list price Manufacturer's recommended retail or wholesale seller's price as shown on manufacturer's price list.

listed building A building of special architectural and/or historic interest listed under the Planning (Listed Buildings and Conservation Areas) Act 1990. (See listing.)

listed building consent The permission required under the Planning (Listed Buildings and Conservation Areas) Act 1990, as amended, from the local planning authority for the demolition, alteration or extension of a building listed as of architectural and/or historic interest.

listed company Company which has obtained permission for its shares to be admitted to the international stock exchange official list.

listed security Security that has a quotation on a recognised stock exchange.

listing The process where buildings are placed on a register of buildings of architectural and/or historic interest maintained under the Planning (Listed Buildings and Conservation Areas) Act 1990. Such buildings are classified as: Grade 1. Those buildings of exceptional interest. Grade II*. Those buildings which are particularly important and of more than special interest, but not in the outstanding class. Grade III. Those buildings of special interest, but which are not sufficiently important to be counted amongst the elite.

litigation The process where parties place a dispute before the courts for settlement.

lives in being, statutory (See statutory lives in being.)

living will Colloquialism known also as an advanced directive. Normally takes the form of a written statement setting out what types of medical treatment the maker of the will does or does not desire to receive in specific circumstances should he be incapable of giving or refusing consent.

Lloyd's Corporation of underwriters and insurance brokers. The underwriters known as 'names' deposit a substantial sum of money with the corporation and accept unlimited liability.

loan capital Capital used to finance an organisation which is subject to payment of interest over the life of the loan.

local authority An organisation responsible for administering local government; a district or county council, a London borough, a parish council or similar defined administrative area.

local government The system whereby the affairs of the community are administered by regional locally elected representatives. They include responsibilities for such matters as education, rating and valuation, refuse collection, town planning, etc.

Local Government and Housing Act 1989 This Act introduced changes in local government housing finance including the ring-fenced Housing Revenue Account (i.e. the HRA may only contain income and expenditure attributable to local authority tenants, that is it cannot be subsidised from other accounts). This Act also restricted the ability of local authorities to use capital receipts (e.g. from right to buy).

Local Government Management Board Non-profit making organisation which exists to promote good practice within local authorities.

Local government ombudsman Function of the ombudsman is to provide independent, impartial and prompt investigation and resolution of complaints caused through maladministration by local authorities in performing their functions.

local housing company Non-profit making company established to take transfer of tenanted local authority housing stock. The company must be a registered social landlord and most are expected to be companies limited by guarantee.

local land charge A binding charge on land recorded in the local land charges register maintained by local authorities.

local plan Plans which are prepared by the local planning authority showing how they would like to see the future development of a particular area. In the determination of applications for planning permission they form an important part of the decision-making process. Those applications conforming to the proposals in the local plan are granted permission and those not refused. There are different types of local plans prepared for various purposes under different planning statutes, variously called development plans, structure plans, local plans, action area plans, etc. The current main form of local plan is the Unitary Development Plan.

local planning authority   The statutory planning authority responsible under the Town and Country Planning Acts for preparing statutory local plans and granting or refusing applications for planning permission. The allocation of planning functions between county and district councils is complicated, but generally speaking county councils are responsible for strategic plan preparation and minerals and district councils responsible for detailed local plans and the determination of most types of applications for planning permission.

Local Valuation Court   A tribunal established to adjudicate on appeals against existing entries on the valuation list, or proposals to amend it. There is a right of appeal from the Local Valuation Court to the Lands Tribunal.

location plan   A plan showing the location of a particular building, normally drawn on a scale of 1 to 1250 or 1 to 2500.

loco standii   Literally 'the right to stand'. It is the right to bring an action or to challenge a decision.

lodger   Person who lives as part of a family and normally shares heating, cooking and other facilities and does not have an exclusive right of occupation of a room or rooms. The Housing Act 1985, as amended, gives secure tenants the right to take in lodgers with the consent of the landlord.

logistics   The management of the flow of information and materials through an organisation.

London Bankers' Clearing House   Central organisation which settles inter-bank indebtedness arising from daily cheque clearings.

London Clearing House   Independent clearing house which provides future and options markets with netting and settlement services. The LCH is owned by six major UK commercial banks and acts as a counter party to every transaction between its members and takes the risk of members defaulting.

London FOX (London Futures and Options Exchange) Commodity exchange which acts as an umbrella organisation for various soft commodity futures market exchanges and which offers futures and options contracts in defined commodities.

London inter-bank bid rate (LIBID) Rate of interest at which banks bid for funds to borrow from each other.

London inter-bank offered rate (LIBOR) Rate of interest at which banks will lend to each other in the inter-bank market.

London International Financial Futures and Options Exchange (LIFFE) Financial futures market which provides facilities for dealing in options and futures contracts including government bonds, stock and share indexes and foreign currencies.

London Stock Exchange London based central market for dealing in freely transferable stock, shares and securities of all types. It produces a daily list of buying and selling prices of the securities which are quoted. These include shares in public companies, government securities, local authority loan stocks and stocks, bonds and securities of foreign companies and governments admitted to the stock exchange listing.

long position Dealer position in which a dealer's holding exceeds sales. The dealer expects prices to rise so enabling a profitable sale of longs.

long-term bond A bond that has more than one year to run before maturity.

Lord Justice of Appeal Title of a judge of the Court of Appeal.

loss adjuster A professionally qualified person responsible for quantifying the losses under an insurance claim normally acting for insurers in an independent capacity.

loss assessor A professional person acting for a claimant in an insurance loss who is responsible for determining the amount of damage incurred and to negotiate the best financial settlement on behalf of the claimant.

loss leader Product or service which is offered for sale by an organisation at a loss in order to attract customers.

lot A property offered for sale, often by auction.

lotting A situation used at auctions where property is divided into parts, each of which is capable of being sold separately with a view to achieving a greater return on the whole than if sold as a whole.

low cost home ownership A colloquial generic term for leasehold schemes for the elderly, shared ownership, do-it-yourself shared ownership, and improvement for sale schemes.

lump sum building contract A building contract in which the contractor agrees to undertake the works for a fixed price.

lunar month A period of 28 days.

M

machine hour rate A method of charging overheads to cost centres where the budgeted or estimated overhead costs attributed to a machine or group of similar machines is divided by the appropriate number of machine hours, i.e. the number of hours which would relate to normal working or full capacity.

magistrate An officer with judicial powers in matters of a minor criminal or civil nature. (See justices of the peace.)

magistrates court The inferior criminal and civil courts. They normally consist of two or more justices of the peace sitting as a court and advised on legal matters by a legally qualified clerk to the justices.

mainstream corporation tax Liability for corporation tax for an accounting period after the relevant advanced corporation tax has been deducted. Payments on account of advance corporation tax are paid when dividends are paid to shareholders. Mainstream corporation tax is the main balance.

maintain A term in a lease covenant meaning to keep substantially in the same condition as when the lease was granted.

maintenance (1) An order following divorce, nullity or judicial separation requiring either party to a marriage to make periodic payments for the maintenance of the other. (2) The necessary periodic work involved to ensure that a building does not fall into a dilapidated state and involves such work as routine cleaning, painting and decorating and ensuring all mechanical services are in good working order.

maintenance and education clauses Clauses in a deed or a will empowering the trustee or trustees to spend the income in the maintenance and education of child beneficiaries.

maintenance contracts Contracts to provide for the inspection and overhaul of servicing.

maintenance period Defects liability period.

maintenance trust fund A fund established for the purposes of maintaining a particular building specified in the trust deed. Normally applied to a single building in multi-occupation, to ensure that funds are available when necessary.

maisonette A dwelling where the living accommodation is situated on two upper levels of a building.

major interest Freehold or leasehold interest. A term exceeding 21 years.

majority Full age, the age at which a person is considered legally competent, now 18 since the Family Law Reform Act 1969.

making good Remedial work involved to existing parts of a building which have been affected by new or additional work.

maladministration Matters which are investigated by the relevant 'ombudsman' for administration following complaints of injustice in consequence of neglect by a public body.

mala fide In bad faith.

male issue Male descendants through the male line only.

malfeasance Undertaking of an unlawful act.

management The running of an organisation or part of an organisation. The term is also used to describe those people who collectively or individually are responsible for running the business or organisation.

management accountant Accountant responsible for the production of management accounts. The major professional body of management accountants within the UK is the Chartered Institute of Management Accountants (CIMA).

management accounting The provision of financial information to the various levels of management within an organisation for the purposes of planning, decision making and monitoring and controlling performance.

117

management agreement   In the housing association context, the situation where a housing association wishes to devolve its responsibility for managing a scheme to another agency. A management agreement is the legal contract which defines the responsibility of the two parties.

Management and Maintenance Allowances   In the housing association context, notional sums of money which will be spent on managing and maintaining the property. The allowances are used in the calculation of the Rent Surplus Fund and to assess the reasonableness of a claim for a Revenue Deficit Grant.

management buyout   Acquisition of an organisation by a team of existing managers. The process often involves substantial institutional backing by venture capital organisations.

management by exception   Control and management of costs, revenues and performance by concentrating on those instances where significant variances occur.

management consultant   Person who specialises in giving advice to organisations on management problems and on ways of improving efficiency and profitability.

management information system   Information system designed to provide financial and quantitative information to management within an organisation.

managing agent   An agent employed by the landlord to undertake a property management function. In the housing association context, an organisation, usually a firm of estate agents, which provides housing management services.

managing director   Company director who usually ranks first in seniority under the chairperson and is normally the chief executive. This individual will be responsible for the day-to-day running of the company and carries ultimate responsibility for management decisions.

managing trustee   A trustee responsible for managing property under the control of a trust.

mandate   An authority given by one person to another to carry out specific duties. A mandate is automatically terminated by the death of the mandator.

map   A drawing which is a scaled representation of the features of a specified area of the earth's surface, drawn to a particular scale. For example, ordnance survey maps at a scale of 1:1250.

Mareva injunction Now known as a freezing injunction. An injunction taking its name from the case Mareva Compania Naviera SA v International Bulkcarriers SA (1980). May be granted to restrain a defendant from transferring abroad any of his assets.

margin The number of percentage points the lender adds to the index rate to calculate the adjustable rate mortgage (ARM) interest rate at each adjustment.

margin of safety The excess of budgeted level of activity over the level of activity required to break even. This is sometimes expressed as a percentage of budgeted or capacity level of activity.

marginal cost The extra cost incurred as a result of the production of one additional unit of production.

marginal costing An approach to costing in which only the variable costs are charged to cost units. The fixed costs are not apportioned to individual units or activities but are met out of the total contribution generated.

Market and Opinion Research International (MORI) Market research organisation which carries out a wide variety of research activities including social and political research and MORI opinion polls.

market capitalisation Calculation of the market value of a company obtained by multiplying the number of its issued shares by their market price.

market forces Forces of supply and demand which in a free market situation determine the quantity available of a particular good or service and the price at which it will be offered. In general a rise in demand will cause both supply and price to increase whilst a rise in supply will cause a fall in price and a drop in demand.

market leader Firm with the largest market share within a particular industry.

market maker A dealer who is willing to buy or sell particular securities at all times.

market penetration The act of entering an established market with a new brand or product.

market price The price paid for a property; the amount of money that must be given or which can be obtained at the market in exchange under the immediate conditions existing at a certain date. To be distinguished from market value.

119

market share The proportion of the total supply of a product controlled by any one firm or company.

market value The highest price estimated in terms of money which a buyer would be warranted in paying and a seller justified in accepting, provided both parties were fully informed, acted intelligently and voluntarily and, further, that all the rights and benefits inherent in or attributable to the property were included in the transfer.

marketable title A title that is free and clear of objectionable liens, clouds or other title defects. A title which enables an owner to sell his property freely to others and which others will accept without objection.

marketing mix Those factors controlled by a company which can influence consumers' buying behaviour. These are often referred to as the four Ps, i.e. product, price, promotion and place.

marriage The voluntary union of one man and one woman for life to the exclusion of all others. The legal requirements concerning a marriage are contained in the Marriage Acts 1949 and 1983. The Acts require that certain formalities must be complied with to constitute a valid marriage. A void marriage is one where the parties have gone through a ceremony but there is lacking some essential ingredient. A voidable marriage is a valid one subsisting until a decree of nullity is pronounced.

marriage consideration Persons within the consideration are husband, wife, issue of the marriage and grandchildren. (See also consideration.)

marriage settlement A conveyance of property for the benefit of the parties to a marriage. These can be varied by the court.

marriage value The potential value which can be achieved by the merger of two or more interests in land.

marriage, will in contemplation of Normally a will is revoked by a testator's marriage but if it appears from the will that at the time it was made the testator was expecting to be married to a particular person and that he intended the will should not be revoked by his marriage, the will will not be revoked by the marriage to that person. See the Wills Act 1837 as substituted by Administration of Justice Act 1982.

master An official of the High Court who decides many interlocutory matters.

master budgets The overall budgets of an organisation built up from a range of individual budgets and comprising the cash budget, the forecast profit and loss account and the forecast balance sheet.

materiality convention Acceptance that there are some transactions or events that are not significant enough for accountants to disclose them.

maternity rights or maternity entitlement The statutory rights of pregnant women and mothers of babies.

matrimonial causes Divorce, nullity and judicial separation suits.

matrimonial home Property where husband and wife have lived together. Under the Matrimonial Homes Act 1983, as amended, a spouse is given protection from eviction of the matrimonial home.

matrix management Management system in which some employees report to more than one manager.

McCarthy Rules The name given to the principles governing the right to compensation for injurious affection where no part of the claimant's land is acquired.

mean A system of averaging used in statistics.

mean deviation A statistical term for the arithmetic mean of the deviation of all the numbers in a set of numbers from their arithmetic mean.

mean price The arithmetic mean of two given prices, e.g. the buying and selling price. Mean price is often referred as the middle market price.

mechanistic organisation Organisational type characterised by clearly specified roles and definitions of rights and obligations within a hierarchical structure.

median A mean in which a set of numbers is arranged in ascending or descending order and the middle number or arithmetic mean of the middle two numbers is taken as the median.

member of a company Shareholder of a company whose name is entered in the register of members.

members' voluntary liquidation The winding up of a company by resolution of the members in circumstances in which the company is solvent.

memorandum of association Document which a company must file with the Registrar of Companies and which is open to public inspection. It must contain the company name, statement that the company is a public company, the address of the registered office, the objects of the company, a statement of limited liability, the amount of the guarantee, the amount of authorised share capital and its distribution.

memorandum of satisfaction Document stating that a mortgage or charge on a property has been repaid.

mental disorder Defined under the Mental Health Act 1983 as an illness or incomplete development of the mind, psychopathic disorder or other disability of the mind.

Mental Health Act 1983 Act which covers all aspects of compulsory hospital admission and subsequent treatment of people with mental illness.

Mental Health Act Commission Government body established to monitor the care of people who are detained to ensure that their rights under the Mental Health Act of 1983 are upheld.

mental illness specific grant Grant from the Department of Health to local authorities for community based mental health projects.

merchant banks Banks which historically specialised in financing foreign trade, but which have recently tended to diversify into hire purchase finance, the granting of interim loans, provision of venture capital, underwriting new issues and managing investment portfolios and unit trusts.

merchantable quality A term used in relation to the sale of goods. There is an implied condition that goods should be as fit for their ordinary purpose as it would be reasonable to expect taking into account any description applied to them, the price and all other relevant circumstances.

merger Combination of two or more organisations into one unit with a view to increasing overall efficiency.

merger of interests The process whereby a superior interest with one or more inferior interests in the same property is amalgamated, thereby amalgamating several titles into one.

messuage A house including gardens, courtyard, orchard and out-buildings.

metric system System of measurement based on the decimal system.

micro-economics Study of economic behaviour at the level of the firm or individual.

middle man Person or organisation which makes a profit by acting as an intermediary between buyers and sellers.

military will or testament A privilege will made by a soldier on active service without those forms which in ordinary cases are required by statute. See the Wills Act 1837 section 11.

minimum lending rate Introduced in 1972 to replace bank rate. It is the minimum rate at which the Bank of England would lend on the discount market.

minimum rent The rent below which a variable rent will not fall.

minimum subscription The amount stated in the prospectus of a new company which the directors consider the minimum that must be raised for the company to be launched successfully.

minimum wage The minimum wage which an employer must pay an employee.

minor Defined under the Family Law Reform Act 1969 as someone under the age of 18 years.

minor interest Third party rights and interests in land which are not registered on the Land Register. They include, inter alia, restrictive covenants, equitable interests, estate contracts, etc. To be binding on allcomers, a minor interest must be protected by an entry on the register of the land affected.

minor tenancy Under the Compulsory Purchase (Vesting Declaration) Act 1981, defined as 'a tenancy for a year, or from year to year, or any lesser interest'.

minority (1) Being below 18 years of age (see majority). (2) The smaller group at an assembly or in a voting procedure.

minority shareholders Shareholders in a company in which the controlling interest is held by another company.

MIRAS Mortgage interest relief at source.

misrepresentation A statement which is false or misleading in fact made by, or on behalf of, one party to a contract to another party to the contract which, although not forming part of the contract, induces another party to enter into the contract.

mitigation Abatement of a loss.

mitigation of loss The situation required of the plaintiff/claimant to take reasonable steps to reduce or avoid loss.

mixed development   A development involving two or more different uses.

mixed economy   Economy in which some goods and services are provided by private enterprise and others by the State.

mixed funding   The use of private alongside public money to finance developments.

mixed hereditament   A hereditament, partly residential, partly commercial, of which the proportion of the rateable value, attributable to the part used as a residential hereditament, exceeds one-eighth.

mobilia sequunter personam   Literally, moveables follow the person. (See moveables.)

mobility allowance   Benefit paid to a person with a physical disability, who is unable to walk. Replaced by the Disability Living Allowance and Disability Working Allowance Act 1991, as amended.

mobility housing   Dwellings designed for occupation by people with a mobility impairment, and who may need to use a wheelchair occasionally.

mobility of labour   Extent to which labour is willing to move from one region or country to another or change from one occupation to another.

model rent review clause   A clause concerning rent reviews, published as a standard in the drafting of leases. The Royal Institution of Chartered Surveyors and the Law Society publish a number of model clauses.

modernisation   Changes undertaken to a building, the purposes of which are to reflect contemporary developments of design, function, decoration, etc.

monetarism   Monetarist economic policies are based on the assumption that price levels are closely related to the quantity of money in circulation and that an increase in the money supply will lead directly to an increase in inflation.

monetary control   The use of the central bank of a country by its government to control the money supply.

money market   UK market for short-term loans consisting of discount houses, banks, the government and accepting houses with the Bank of England acting as lender of last resort.

money supply   The quantity of money in circulation. There are a number of measures of the money supply in the UK ranging from M0 to M5. The narrowest of these M0 refers to notes and coins in circulation. The broadest definition M5 refers to M0 plus private sector current and deposit accounts, private sector bank deposits plus holdings of money market instruments such as treasury bills, foreign currency bank deposits and building society deposits.

moneylender   Defined in the Moneylenders Acts 1900 and 1927 (repealed by the Consumer Credit Act 1974) as any person whose business was that of moneylending, or who advertised or announced himself or held himself out in any way as carrying on that business, but not including pawn brokers, friendly societies, bankers or bodies exempted by the Department of Trade and Industry. Moneylenders are normally regulated by the Consumer Credit Act 1974, as amended.

monitoring   In the context of housing associations, the regular surveillance of the work of each registered housing association by the Housing Corporation.

monopoly   A market in which there is only a single seller or producer.

monopsony   A market in which there are many sellers but only one buyer.

Monthly Digest of Statistics   Monthly publication of the UK Central Statistical Office providing statistical information on industry, national income and UK population trends.

mortgage   An equitable interest in freehold or leasehold property which is conveyed as security for a loan with provision for redemption on repayment of the loan. The mortgagee (lender) has powers of recovery in the event of default by the mortgagor (borrower). For residential properties, mortgages are normally either repayment or endowment. Repayment mortgages are where the borrower throughout the term repays interest and capital. At the end of the term the mortgage is paid off and discharged. For endowment mortgages the borrower also takes out a life policy in the same sum as the loan. No capital is paid during the term. On maturity of the policy the capital balance is repaid and the mortgage discharged.

mortgage broker   A professional that helps consumers through the loan selection, processing and closing of a mortgage loan. Most mortgage brokers have access to a wide range of mortgage products through many mortgage lenders. Mortgage brokers are paid a fee by the borrower when a suitable mortgage is found and closed.

125

mortgage commitment A written notice from the bank or other lending institution saying it will advance mortgage funds in a specified amount to enable a buyer to purchase a house.

mortgagee The lender of money or the receiver of the mortgage document.

mortgagees remedies If a borrower defaults, the remedies available are as follows: firstly foreclose, secondly exercise the power of sale, thirdly sue on a personal covenant, fourthly appoint a receiver, and fifthly take possession of the property.

mortgagor The person who mortgages his property as security for the mortgage debt, the borrower.

mortis causa donatio (See *donatio mortis causa*.)

Move On Grant In the housing association context, a Housing Corporation Grant paid per bed space on self-contained units developed under the Housing Act 1988, as amended. The arrangements are for people who have moved out of institutions.

moveables Goods, furniture, etc. which may be moved from place to place.

multiple agency The appointment of two or more agents to sell or let property. The successful agent is entitled to commission which does not have to be shared with other retained agents.

multi-storey A building of five or more storeys.

municipal law The laws affecting the administration and implementation of local government.

mutatis mutandis With the necessary changes.

mutual option Option available to either party to a contract in the event of a specific happening.

mutual wills An arrangement in which two or more persons draw up a will conferring reciprocal benefits, that is in favour of each other. These wills may be revoked during the lifetime of all testators but a court will enforce the terms of the mutual wills following the death of any of the testators.

N

NACAB Abbreviation for National Association of Citizens' Advice Bureaux.

NACRO Abbreviation for National Association for the Care and Resettlement of Offenders.

naked trust (See bare trust.)

name and arms clause A clause in a will that specifies that a beneficiary may only benefit if she uses a specified last name and, where appropriate, coat of arms.

national assistance A form of benefit introduced by the National Assistance Act 1948 and superseded by supplementary benefit introduced under the Supplementary Benefit Act 1966, as amended. This in turn has been superseded by income support.

National Association of Estate Agents An organisation established in 1962 to represent and protect the interests of practising estate agents.

National Audit Office (NAO) Central government organisation responsible for external audit of other central government departments and agencies. The National Audit Office is accountable to Parliament through the Public Accounts Committee.

National Code of Local Government Conduct A statutory code which governs the conduct of local authority members.

National Council for Voluntary Organisations (NCVO) Representative body for national charities and other non-profit making organisations.

National Federation of Housing Associations The central organisation which represents housing associations in England.

National Health Service and Community Care Act 1990 Legislation which introduced major changes to the way health and community services were managed. Under the Act local authorities were given clear lead responsibility for community care and the resources for funding residential care were transferred from social security to social services.

National Health Service The service introduced by the National Health Service Act 1946, which is responsible for providing comprehensive health care to residents of the United Kingdom irrespective of income.

National Health Trust A body responsible for the ownership and management of hospitals and other establishments which were previously administered by regional, district or special health authorities.

National Housing Federation Central representative negotiating and advisory body for registered social landlords and other non-profit housing organisations in England.

National Insurance Contribution Payment made by those with earned income which contributes to the national insurance fund from which benefits are paid.

National Insurance The means by which various social benefits are funded and supported. Claimants for benefits may appeal against the decision of an adjudication officer to a social security appeal tribunal, or disability appeal tribunal. Further appeal on a point of law is to the Social Security Commissioner.

National Loan Fund In housing association context, the source of government loans to the Housing Corporation.

National Savings Bank Savings bank operated by the Department of National Savings.

National Savings Bond Long-term investment offered by the Department of National Savings.

National Savings Certificates Government security on which neither income tax nor capital gains tax is payable.

National Tenants Resource Centre National charity providing training and other support to social housing tenant groups.

National Trust A charitable trust established under the National Trust Act 1907 responsible for the preservation and conservation of places of historic and/or architectural interest and natural beauty.

nationalisation Process of bringing privately owned assets into the ownership of the State.

natural light Light obtained from the sky used to illuminate rooms. The degree of illumination being dependent upon the amount of glazed area serving the room. (See daylight factor.)

natural right of support A right at common law which prevents the owner of an adjoining property from doing something which endangers the stability of the surface and the sub-strata of land adjoining the right, or neglecting to do something which is needed to maintain the stability. No natural right of support to a building exists at law.

negative development value The situation where existing use value is greater than the value of the property following development or redevelopment.

negative value The situation in which the value of an asset which is a liability, can be disposed of only by means of a payment to the purchaser.

negligence A breach of a duty of care which results in damage. Actionable in law as a tort.

negotiation The process whereby two, or more, parties aim to reach a common agreement. May be verbal or written.

negotiator A person who undertakes negotiations often as an agent on behalf of the principal.

nemo est heres viventis Literally 'no one is the heir of the living person'. A person's heir cannot be ascertained until that person's death. (See heir apparent.)

net book value Cost of an asset less its accumulated depreciation to date. This is also known as the written down value.

net cash flow Net cash flow or outflow of cash resulting from proceeding with an investment project.

net current assets Working or current capital calculated by subtracting current liabilities from current assets.

net dividend Dividend paid by a company to its shareholders after excluding the tax credit received by shareholders.

net estate The property that a deceased person can dispose of by will, less funeral and administrative expenses.

net income Income of a person or organisation after the deduction of appropriate expenses.

net internal area The measured usable space within a building. Measurements are taken to the internal finishes, but exclude toilets, lift and plant room, stairs and lift wells, common entrance halls, lobbies and corridors, internal structural walls and car-parking areas. Often referred to as effective floor area.

net present value Value obtained by discounting all cash outflows and inflows attributable to a capital investment project by a chosen percentage.

net profit   Profit of an organisation when all receipts and expenses have been taken into account.

net realisable value   Sales value of the stock of an organisation less any costs which would be incurred in selling them.

net rent   Rent less outgoings.

new towns   Large settlements newly created by development corporations established under the New Town Act 1946, as amended.

next of kin   Generally one's closest blood relations.

NHF   Abbreviation for National Housing Federation.

NHS Executive   Section of the Department of Health which is responsible for the management of the National Health Service.

nominated sub-contractor   A firm that is nominated by the architect and which the contractor must employ to carry out a specified part of the building work which is usually specialist in nature, such as heating and ventilating.

nominated supplier   A firm that is nominated by the architect and from whom the contractor must purchase certain building items such as windows, doors and roof tiles.

nomination   A direction to a person holding funds that they should be paid to a named person on the death of the person issuing the direction.

nonage   The absence of full age.

non-conforming use   The use of land and buildings which do not conform with the local planning authority's allocated use for an area. Such uses were normally in existence before statutory planning controls were introduced.

non-contributory pension   Pension scheme which requires no contribution from an employee during their employment.

non-departmental public body   A public body which is not directly under the supervision of a minister. This is also known as a quango.

non-executive director   Director of a company who has a seat on the board but is not involved in the day-to-day management of the organisation.

non-recourse loan A loan where the terms provide that the only security to the lender is the property offered as collateral.

non-recurring expense Expenditure which is usually not repeated. For example, works required to meet a statutory requirement.

non-taxable income Income which is not liable to income tax.

notes to the accounts Information supporting that given within a company's financial statement. Notes detailing capital assets, investment, share capital and reserves are required to be given by law. The notes must also give details of the accounting principles which have been followed in the preparation of the accounts.

notice to quit A periodic tenancy runs indefinitely unless either the landlord or tenant serves a notice to quit. It must specify the correct date for termination of the tenancy which has to be an anniversary date.

notice to treat A notice served on the owners, mortgagees and lessees by a public authority when using its compulsory purchase powers. The notice provides details of the property to be acquired, demands details of ownership and recipients' claim for compensation. It indicates that the acquiring authority is willing to treat for the purchase of the land.

nugatory Useless, invalid.

nuisance Legally there are two types: firstly a public or common nuisance; and secondly a private nuisance. A public or common nuisance is an act which interferes with the enjoyment of a right which everyone is entitled to, such as the right to travel, the right to fresh air, etc. A public nuisance is a crime. A private nuisance is any wrongful disturbance or interference with a person's use or enjoyment of land or an act allowing the escape of deleterious things on to another's land, e.g. water, smoke, smell, fumes, etc.

null and void Invalid, without force.

numbered account A bank account which is identified only by a number.

nuncupative will An oral (rather than written) testament/will. These were abolished by s 9 Wills Act 1837 except in the case of soldiers and sailors on active service. See the Wills (Soldiers and Sailors) Act 1918.

NVQ Abbreviation for National Vocational Qualification.

O

oath A religious statement appealing to God to witness that what is said is the truth.

obiter dictum Statement of a judge on a point not directly relevant to his decision and therefore not strictly of authority.

objects of a power Where property is settled subject to a power among a restricted class, the members of the class are known as the 'objects of the power'.

obsolescence Impairment of desirability and usefulness brought about by changes in the art, design or process or from external influencing circumstances that make a property less desirable and valuable for a continuity or use.

occupation The act of taking physical control or possession of land.

occupational pension scheme A pension scheme open to the employees of a particular trade or profession or working for a particular organisation.

occupiers liability (See dangerous premises.)

offer A promise by one party to do a specified deed as the other party in turn performs a specific deed.

officer of a company Person acting in an official capacity on behalf of a company. This will include directors, managers, the company secretary and in some cases auditors and solicitors.

official custodian for charities A person who acts in the role of trustee for charitable trusts and in whom charity property is vested by court order.

official list A list of all the securities traded on the London Stock Exchange at the end of each day's dealings giving current prices of shares, etc.

official receiver A person appointed by the Secretary of State under the Charities Act 1960 in matters concerning bankruptcy.

official referee Former name for a judge who tries technical cases in the Queen's Bench Division of the High Court.

official solicitor An officer of the Supreme Court who acts for persons under a disability.

offshore fund   Fund which is held outside the country of residence of the holder.

Old Lady of Threadneedle Street   Vernacular name for the Bank of England.

oligopoly   A market with many buyers and few sellers.

ombudsman   Popular name for the Parliamentary or Local Government Commissioner who is the public officer responsible for investigating and making recommendations concerning mal-administration in central or local government.

on-costs   Additional costs beyond basic costs. Costs which include professional fees, legal charges, insurance and finance.

open contract   A contract for the sale of land which simply states the names of the parties, the price and a description of the land. It leaves statute and common law to determine any necessary terms.

open market value   The best price which could potentially be achieved for an interest in property at the valuation date. The Royal Institution of Chartered Surveyors Guidance Notes on the Valuation of Assets define it as 'the best price which might reasonably be expected to be obtained for an interest in the property at the date of valuation, assuming firstly a willing seller, secondly a reasonable period in which to negotiate the sale, thirdly that values will remain static during that period, fourthly that the property will be freely exposed to the market, and fifthly that no account will be taken of any higher price that might be paid by a person with a special interest'.

operating leverage (operational gearing)   The relative proportion of fixed and variable costs within an organisation's cost structure.

operating statement   Internal management control document usually produced at regular intervals which reports actual costs and revenues compared with the budget.

operative part   The part of a deed in which the object of the deed is effected. (See recitals.)

operative words   The exact wording through which the object of a document is achieved.

opportunity costs   The value of a benefit sacrificed in favour of an alternative course of action.

option   A contractual right enabling one of the parties, if they wish, and if the circumstances are available under the terms of the contract, to exercise a right to do something, or require the other party to do something in the future.

order   A direction of a court.

order of discharge   An order made by a court of bankruptcy, which releases a bankrupt from his debts.

ordinary share   Unit of the share capital of a company.

ordnance bench mark   Reference points indicated on an ordnance survey map which indicate the height or depth below sea level. (See ordnance datum.)

ordnance datum   The point at which the heights and depths of Great Britain are established and from which bench marks are determined. It is derived from mean sea level readings at Newlyn in Cornwall. (See ordnance survey.)

ordnance survey   An organisation established in 1791 to produce maps for military purposes. It is now a government agency responsible primarily for producing topographical maps.

organic organisation   Organisation with relatively little reliance on authority, decentralised control and decision making and roles and tasks which are constantly redefined.

organisational psychology   Application of psychological concepts and practices to the work environment.

organisational behaviour   Study of people's behaviour, attitudes and performance within an organisational environment.

organisational culture   The predominant values, customs and norms within an organisation. Sometimes referred to as the personality of the organisation.

organisational design   The structure adopted by an organisation in order to best achieve organisational aims and objectives.

orientation   The positioning of a building in relation to North, South, East and West.

outbuilding   A small building within the boundary of a main building but not part of it, such as a shed or store.

outgoings   Sums of money paid by the owner of an interest in property. For example, rates, insurance, etc. These are normally expressed on a yearly basis.

outhouse   (See outbuilding.)

outline planning permission The formal consent granted by a local planning authority in response to an application for outline planning permission, subject to the grant of reserved matters usually concerning design, materials and external appearance.

outsourcing The buying in of components, finished products and services from external suppliers.

over In conveyancing a gift or limitation over means one which is to come into being on the determination of some prior interest.

over capacity Situation where capacity is greater than that needed to meet market demand.

over capitalisation Situation in which an organisation has capital in excess of its current needs.

overhead absorption rate Basis on which indirect costs are charged to individual products or services.

overriding interests Interests in land which are set out in the Land Registration Act 1925, s70(1), and bind the buyer of the land irrespective of whether he has notice of them. They include, inter alia, legal easements, rights of occupants and local land charges.

oversale A wire or cable crossing land without support on the land.

owner One who holds title to and conveys the right to use and occupy a property under agreement.

ownership The right to exclusive use, possession or disposal of property. It may be absolute, where the owner may freely use or dispose of the property, or restricted, for example joint ownership.

ownership, beneficial The right of enjoyment of property, as opposed to legal ownership.

P

paid up capital The amount of money that the shareholders of a company have paid to the company for their fully paid shares.

panel solicitors In the housing association context, firms of solicitors approved by the Housing Corporation to act in conveyancing for both the Housing Association and the Housing Corporation under a common code and scale of fees.

par value The nominal or face value of a share or other security.

parage Equity of dignity, blood or name.

parcel A separately identifiable area of land usually in one ownership. The term is often used in association with conveyancing.

parent Father or mother of a child. See Family Law Reform Act 1987 section 1.

parental responsibility The rights, duties and responsibilities and authority which by law a parent has in relation to their children and property. See Children Act 1989.

pari passu In proportion; on an equal footing.

Parker Morris Standards Standards for housing associations' and local authorities' new build schemes established by the government in 1967. They are now superseded by the design criteria published by the Housing Corporation.

parking ratio The ratio of a number of car parking spaces to the amount of accommodation made available for a particular building or development project.

Parliamentary Commissioner for Administration Ombudsman responsible for investigation of complaints referred to him by an MP from a member of the public concerning maladministration by government departments and public bodies.

parole evidence The legal rule which prevents previous oral or written negotiations to a signed contract from changing the contract.

part owners Persons who have a share or interest in a thing.

part possession The situation where part of the property is occupied by the owner, or is vacant, with the remainder being subject to one or more tenancies.

Part Three Homes In the housing association context, residential accommodation for people in need of care and support for reason of age, infirmity or physical disability. These homes were first provided under Part 3 of the National Assistance Act 1948 and are now required to be registered under the Registered Homes Act 1984, if they provide board and personal care.

partial completion certificate A certificate issued by an architect or surveyor to a building contract where the employer wishes to take possession of a completed part of a building.

partial intestacy Arises when the deceased disposes of some, but not all, of the beneficial interest in her property by will.

participators Parties to a joint venture who share the risks and profits or losses of the venture. The benefits and obligations of the parties to the venture would depend on the specific terms of the agreement entered into.

particulars Details of some allegation pleaded in an action. If a pleading is insufficient the opponent may ask for further and better particulars.

particulars delivered Notification to the Inland Revenue at the time when stamp duty is payable on the transfer of a title in land (or to HM Land Registry as agents of the Inland Revenue where a duty is payable).

parties Persons who voluntarily take part together in anything, e.g. parties to a deed.

partnership Association of more than two people formed for the purpose of carrying on a business. Partnerships are regulated by the Partnership Act (1890). As a partnership does not have a legal personality partners are liable for the debts of the firm.

party wall A wall belonging to more than one owner. Generally tenancy in common of a wall is assumed. Under the Law of Property Act 1925 section 38(1), as amended, a party wall is deemed to be severed vertically between the respective owners, each of whom has the requisite rights of support and use over the rest of the wall.

party wall agreement An agreement indicating the respective rights of the owners of properties divided by a party wall. It is an agreement enforceable against purchasers when registered as a land charge.

passing rent Existing rent which is payable under the terms of a lease or tenancy agreement.

passive trust (See barc trust.)

patent Granting of an exclusive right to make, use or sell an invention during the period that the patent remains in force. In the UK patents are granted by the Crown through the Patent Office which is part of the Department of Trade and Industry.

pavement A pathway provided at the side of a road for the use of pedestrians.

pay as you earn (PAYE) System of income tax collection where tax is deducted from wages and salaries by an employer before payment.

payback period The number of years which will elapse before outlay on a capital investment project is recovered.

payment in kind A payment which is made in goods or services rather than cash. This often takes the form of discounts or allowances against goods and services.

payment on account Part payment of a debt.

payroll tax Form of business taxation based on the total amount of an organisation's payroll.

pecuniary legacy A gift of cash made through the instrument of a will. (See legacy.)

penal rent A vastly higher payment due during a period of default by a tenant to honour the obligations to pay rent at the proper time.

penalty A sum provided by a contract as payable in the event of a breach, which sum is deemed not to be liquidated damages.

pendente lite, **administration** An administrator appointed by the court where there is a dispute as to the validity of a will. See Supreme Court Act 1981, s 117.

penny shares Securities with a very low market price traded on a stock exchange.

pension Regular payments made to a retired person.

pension funds Private and state pension contributions invested to provide the funds from which pensions are paid.

penthouse A building constructed on the flat roof of a main building for living purposes, such as an apartment.

PEP Abbreviation for Personal Equity Plan.

peppercorn rent A nominal amount of money, or token rent, payable to the landlord. It is often associated with payment of a premium to the landlord by the tenant.

perfect rights A right recognised and enforced by a legal system.

perfect trusts (See executed trust.)

perfection of gift The properly constituted transfer of ownership of property. (See gift, imperfect.)

performance appraisal Evaluation of the performance of an employee.

performance bond A guarantee that the provider of goods and services will meet the terms of the contract and will pay compensation. In the housing association context, a performance bond may be guaranteed by the Housing Corporation.

performance indicator In the housing association context, the Housing Corporation requires registered housing associations to use and publish performance indicators and arrange housing management finance and development activities.

performance specification A detailed description of the required standards of a material or component.

peripatetic wardens Staff who provide support services to older people living in groups of dwellings which are not part of a sheltered housing scheme.

permitted use A colloquial term meaning: (1) The use of land or buildings for which planning permission has been granted. (2) The use of land or buildings which is deemed to be granted under the general development order.

perpetual injunction An injunction that is granted after the hearing of an action and cannot be dissolved except by appeal.

perpetual inventory Method of continuous stock control in which an account is kept on a continuous basis for each item of stock.

perpetual trusts, rule against (See inalienability, rule against.)

perpetuities, rule against Rule preventing the controlling of the devolution of an estate beyond the period allowed by law. See Perpetuities and Accumulations Act 1964, s 1.

perpetuities rule, exceptions to The rule does not apply in a number of circumstances, e.g. a gift to charity followed by a gift over to another charity.

perpetuity Without a time limit.

personal allowance Sum deducted from taxable income to allow for personal circumstances.

Personal Equity Plan (PEP) UK government sponsored scheme where individuals are able to obtain tax benefits on a savings scheme geared to investment for at least one year.

personal property Anything in ownership other than land, which can include a leasehold interest.

personal representative The executor or administrator of the estate of a deceased person, appointed by the terms of the will or the rules of intestacy.

personalty (See personal property.)

personnel management Part of management which is concerned with people at work and their relationship within an organisation.

presumption of death The presumption that a person has died if she has not been heard of for seven years.

petition A document used to begin certain civil actions such as divorce or a winding up.

petty cash Cash kept within an organisation for the purpose of meeting small incidental expenses on a day-to-day basis.

physical factors Defined under s 1(2) of the Land Compensation Act 1973 as noise, vibration, smell, fumes, smoke, artificial lighting and the discharge onto land of any solid or liquid substance. Compensation may be payable where there is a diminution in value by any of these physical factors created by the use of physical works.

physical life The term during which it is assumed a building is capable of occupation and is capable of meeting accepted standards and statutory requirements.

pilot study Small-scale study conducted as a trial in order to eliminate problems before a full study is undertaken.

pink land Land shown as pink on plans for clearance areas which contains dwellings unfit for human habitation.

plan A scale drawing of the layout of, and construction of, an existing or proposed building, in the horizontal plane. Also a drawing of a relatively small area of land indicating its boundaries, buildings, services, etc.

planner A colloquial term meaning: (1) A Town Planner being a professional person who engages in the practice of Town and Country Planning, usually a Corporate Member of the Royal Town Planning Institute. (2) A person who is engaged in the management of the production of a building project. Sometimes called a planning engineer.

planning By virtue of the Town and Country Planning Act 1947, as amended, the statutory control of development which includes both buildings and use of land.

Planning (Listed Buildings and Conservation Areas) Act 1990 A consolidating Act relating to most of the provisions relevant to heritage properties, although those relating to ancient monuments are still separate in the Ancient Monuments and Archaeological Areas Act 1979.

Planning and Compensation Act 1991 Introduced four significant changes to the planning framework, the first to development control, where a 'plan-led' system was adopted; the second was to make the adoption of district-wide local plans mandatory; the third was to abolish the requirement for central approval of structure plans; and the fourth was to introduce a mandatory requirement for counties to produce mineral plans and waste plans for the whole of their areas.

planning appeal The process whereby a person who has been refused planning permission or an enforcement notice appeals to the Secretary of State for a more favourable outcome. A Planning and Housing Inspector is appointed by the Secretary of State to review the decision of the local planning authority and issues a reasoned letter informing the appellant of his decision.

planning application The making of an application to the local planning authority for permission to undertake development. The application can be for either outline or detailed permission; the former for permission in principle and the latter for permission of the whole scheme.

planning blight The diminution in the value of land adversely affected by a planning proposal contained in a development for its acquisition at some future date by a public authority. Owner-occupiers in certain circumstances may serve notice on the local authority requiring it to purchase the land.

planning brief Planning policies and guidelines indicating the future development pattern of an area of land. Usually a term referring to a document issued by a local planning authority.

planning consent A colloquial term meaning that planning permission has been granted.

planning obligation Replaces the term planning agreement. A person interested in developing a piece of land may by agreement or otherwise enter into an obligation: (1) Restricting use of the land in a particular way. (2) Requiring specified operations or activities to be carried out. (3) Requiring the land to be used in a specified way. (4) Requiring a sum or sums to be paid to the authority.

planning permission The formal consent of the local planning authority in response to the submission of a planning application. Granted either unconditionally or with conditions attached. Permissions are normally in response to applications for outline or detailed permission.

PLC Abbreviation for Public Limited Company.

pleading A written statement of a party in a case to a civil action.

pointe gaurde A concept in the assessment of compensation where land is acquired compulsorily. That any increase or decrease in the value of the land taken resulting from the scheme underlying the acquisition, shall be ignored.

points of claim, defence The title of pleadings in an arbitration.

political objects These objects are not recognised as charitable objects and include advancing the interest of a political party.

portfolio A collection of investment properties held in one ownership.

portion Gift of property personal or real made to a child by a father or one *in loco parentis* in order to establish the child in life or to make a permanent provision for her.

Positive action training in housing (PATH) Schemes for training people from minority ethnic groups to equip them for jobs with registered social landlords or local authorities.

possession, interest in (See interest in possession.)

possessory title (1) An ownership claim to land based on inconclusive or non-existent evidence and which, within a prescribed period, is challengeable by a person who has a stronger claim. (2) Regarding registered land, the title which is subject to some qualification, or exception as stated in the register.

poundage The amount payable by a ratepayer for every pound of rateable value.

poverty trap Situation in which people with low incomes who are means tested for benefits find that any increase in their income results in a similar cut in benefit.

power An authority, as opposed to duty, to dispose of property, real or personal not your own. Where an action goes beyond the scope of the power it is invalid and can be challenged in the courts.

power coupled with an interest The same person holds a power to perform some act combined with an interest in the subject matter of the act.

power of attorney A written instrument authorising a person to act as the agent of the person granting it, and a general power authorising the agent to act generally on behalf of the principle. A special power limits the agent to a particular or specific act such as: a landowner may grant an agent special power of attorney to convey a single and specific parcel of property. Under the provisions of a general power of attorney, the agent having the power may convey any or all property of the principal granting the general power of attorney. An enduring power of attorney cannot be revoked in the event of the subsequent incapacity of the grantor.

power, capricious A power which is void as 'it negatives any sensible consideration by the trustees of the exercise of the power': *Re Manisty's Settlement [1974] Ch 17.*

practical completion certificate A certificate issued by an architect or surveyor, in relation to a building contract at the time when the building is complete in virtually all respects, and ready for occupation. The certificate authorises the release of an agreed sum of money and any retention money.

preamble An introduction to a section or part of a document such as a bill of quantities.

precatory trust A trust arising as a result of the use of precatory words.

precatory words Expressions of hope or desire which do not place an imperative obligation on the possible 'trustee'.

precedent condition (See condition precedent.)

precept The amount of money demanded from rating authorities by a precepting authority, such as a parish council. The amount is normally expressed as part of the rates and is required to cover the cost of running the precepting authority.

precepting authority The authority which precepts.

pre-contract deposit An amount of money paid in advance of the contract as an expression of good faith, and usually held by a third party; for example, a solicitor. A deposit normally acts as part-payment on completion of the contract. It does not create a legally binding agreement in the meanwhile.

predecessor One from whom a benefit is derived of succession to property.

pre-emption The right of first refusal.

preference shares Normally fixed interest shares whose holders receive dividends in preference to ordinary shareholders but after debenture and loan stock holders.

pre-funding Finance provided on a long-term basis for a development project arranged before commencement of the building works.

pre-let An enforceable agreement in law for a letting which takes effect at a future date.

preliminaries The introductory section to a bill of quantities.

premium (1) An amount paid by an actual, or prospective, lessee to a lessor, usually the rent is reduced in return for the consideration. (2) A colloquial term meaning the price paid for the purchase of a leasehold interest. (3) Sums payable to an insurer by the insured.

premium rent A rent considered to be above a reasonable level which would be expected to be commanded in the open market on normal terms.

prescription A claim to some right, based upon long user.

present value The future worth of property discounted to its present day value.

presumption An assumption that is made until evidence proves otherwise.

presumption of advancement (See advancement.)

presumption of survivorship (See commorientes.)

presumptive heir (See heir, presumptive.)

price competition Competition based on the price at which a good or service is offered.

price control Intervention by the government to keep prices stable.

price: earnings ratio An investor's ratio which is calculated as the current stock market price per share divided by the most recent available earnings per share.

price leadership The setting of the price of a good or service by a dominant firm in an industry.

price payable The amount paid by the lessee of a dwelling house who is enfranchised.

prices and incomes policy A government policy which aims to curb inflation by imposing wage restraint and price controls.

primary co-operative In the housing association movement, a housing co-operative; a meaning which distinguishes the organisation providing housing from a secondary co-operative which supplies services to it.

primary residence The main or permanent residency of an owner or lessee.

prime cost A sum specified in a bill of quantities.

prime cost contract A building contract where the contractor's payment is calculated on the actual cost of materials, labour and plant, etc. plus an agreed amount or percentage for administration, overheads, and profits.

prime costs The direct costs of production generally comprising direct material costs, direct labour costs and other direct production expenses.

prime property Best property available when viewed from an individual person's perspective.

prime tenant A tenant with a high reputation for meeting their obligation under the terms of a lease. May be existing or potential tenant.

primogeniture The system of inheritance, abolished under the Administration of Estates Act 1925 Part IV whereby preference was given to the eldest son and his issue.

principal (1) The person who actually commits the crime. (2) The person on whose behalf the agent acts. (3) A sum of money lent or invested. (4) An heirloom.

Priority Estates Project Project originally sponsored by the Department of Environment which advises local authorities and

145

registered social landlords on how to deal with housing estates with a history of management difficulties.

priority investment areas Rural or urban areas of housing stress which have priority for public investment. They are determined by the regional offices of the Housing Corporation.

priority need A term used to define the categories of statutory homeless households for which a local authority must provide accommodation, provided they are not intentionally homeless.

Private Finance Initiative (PFI) Government programme to finance hospitals, roads, prisons, etc. with private lending.

private sewer A pipe situated below ground level which is used for disposing of foul and surface water from drains of a group of buildings and maintained by the owners.

private tender A tender restricted to specified named parties.

private treaty The process for the disposal of real property whereby negotiations are carried out between the vendor and prospective purchaser privately. Usually without any limit on the time within which they must come to an agreement before contracts are exchanged.

privatisation Act of selling publicly owned company to the private sector.

privileged will The right of a person on active military service to make an informal will, i.e. one not made following the normal formalities, which will be held valid.

Privy Council The judicial committee of the Privy Council is the final court of appeal for commonwealth countries.

probability The likelihood that a particular event or circumstance will occur.

probability tree Tree-like chart which sets out assessed probabilities associated with pessimistic and optimistic variations from the predicted levels of each variable. This leads to the probabilities of the most pessimistic and most optimistic outcomes for the project as a whole.

probate A court document evidencing the authority of the executor of a will. The court must 'grant probate' when a will is contested.

probate action Court proceedings for the granting or revocation of probate and to determine whether or not a will is valid.

Probate Court A court established to which all non-contentious probate matters are assigned. The Chancery Division now governs its work.

probate duty Formerly called stamp duty on grant of probate on the value of an estate of a person who died intestate. Replaced by estate duty.

probate registry The office which deals with the issue of grant of probate.

probate rules Rules of court concerned with the regulation of practice with respect to non-contentious probate business.

probationary tenancy (See introductory tenancy.)

process Managed set of work activities with known input designed to produce a specific output.

product differentiation Distinction between products which fulfil the same purpose but may be made by different producers and therefore are in competition for sales.

product diversification The strategy of marketing new products to a new set of consumers.

product enhancement The development of a new and improved version of an existing product in order to extend the lifecycle of a product.

product lifecycle The pattern of a product's sales and profitability over its lifetime.

production overhead costs Indirect costs of production, usually labour and materials not unique to one product or service and indirect expenses of running plant and production premises.

professional indemnity insurance Insurance taken out by professional advisers, e.g. architects and surveyors against actions in negligence, or other relevant matters.

professional witness Individual who gives evidence in court either as part of their job or because of their professional expertise.

profit and loss account The summary of an enterprise's transactions over a stated period which shows revenue generated with the related costs plus the profit or loss for the period. It may also show the appropriation of profit before tax, dividends and reserves.

profit centre   A responsibility centre where the manager is responsible for costs and revenues and therefore for the profit obtained within a given level of capital investment.

profit variance   The difference between standard profit on budgeted sales volume and the actual profit for a specific period.

profitability index   Statistic used in investment appraisal calculated as the present value of cash inflows from a project divided by the present value of cash outflows from a project.

prohibition notice   A notice served by an inspector appointed under the Health and Safety at Work, etc. Act 1974, as amended, which states in the opinion of the inspector, that certain activities at work involve, or will involve, a risk of serious personal injury. It specifies the matters giving rise to the risk and indicates the alleged contravention of statutory provisions and directs that the activities concerned shall not be carried out unless the matters specified are rectified. The notice is of immediate effect and there is a right of appeal to an Employment Tribunal.

project management   The management of a building or development project which involves the role of co-ordinating budgets, advice from professionals and generally ensuring that the project is completed in accordance with the client's stated requirements.

promissory estoppel   Applies when one party promises not to enforce her rights under a contract. (See also estoppel.)

proper valuation   Under the Insurance Companies Regulations 1981 is defined as 'in relation to land a valuation by a qualified valuer, not more than three years before the relevant date, which determines the amount which would be realised at the time of the valuation on an open market sale of the land, free from any mortgage or charge'.

property   The rights of ownership. The right to use, possess, enjoy, and dispose of a thing in every legal way and to exclude everyone else from interfering with these rights. Property is generally classified into two groups: personal property and real property.

property management   The arrangement of functions including collection of rents, payment of outgoings, maintenance, provision of services, etc. The precise duties between the landlord and tenant will be specified in the terms of the lease.

property management agreement   The contract between an owner and property manager. The precise terms of the agreement being specified in the contract.

proposed dividend Dividend which has been approved by the directors of a company but not yet paid.

proprietary estoppel A doctrine under which the court can grant a remedy when an owner of land has led another to act detrimentally in the belief that rights over land would be acquired. (See estoppel.)

prospectus Document which gives details of a new share issue and invites the public to buy shares or debentures in the company.

protected shorthold tenancy A tenancy where the landlord has an extra ground for claiming possession against a tenant provided it is a shorthold tenancy.

protected tenancy By virtue of the Rent Act 1977, as amended, a tenancy which is rent controlled and there is security of tenure.

Protection from Eviction Act 1977 Gives the right not to be evicted without a court order to virtually all residential occupiers. Unlawful eviction is a criminal offence under the Act. It is also a criminal offence to harass a tenant if the landlord knows that the conduct is likely to cause the tenant to leave the premises.

protective trust A trust for the lifetime of a beneficiary or until a specified event occurs whereupon the trust income will be applied at the absolute discretion of trustees. See Trustee Act s 33.

proving a will Obtaining probate of a will.

provision for depreciation Sum credited to an account of an organisation to allow for the depreciation of a fixed asset.

provisional sum A sum of money included in a bill of quantities to cover the cost of unexpected work.

prudence convention Term used within accountancy to describe the convention of using the lowest of all reasonable values of an asset and of not anticipating revenue or profits.

PSBR Abbreviation for public sector borrowing requirement.

public authorities Organisations exercising functions for the public benefit.

public benefit Beneficial effects on the public at large, or an identifiable and significant group. Charitable bodies must show that their activities confer a public benefit.

public company Company whose shares are available to the public through a stock exchange.

149

public corporation State owned organisation which provides a national service or runs a nationalised industry.

Public Expenditure Survey Committee A committee of ministers and senior civil servants of the spending departments which reviews government spending.

public house Premises licensed for the sale of intoxicating liquor for consumption on the premises.

public limited company (plc) Company registered under the Companies Acts as a public company. It must end its title with the initials plc and have a minimum authorised share capital.

public nuisance (See nuisance.)

public sector borrowing requirement The amount by which UK government expenditure exceeds its income.

public sewer A pipe situated below ground level used for disposing foul and surface water from drains and private sewers of buildings and is the responsibility of the water authority.

public tender/open tender A tender which is available to any member of the public to submit a tender provided they meet any terms specified in the tender document.

public trust A trust which has as its object the promotion of the public welfare, as opposed to a private trust, which is for the benefit of an individual or class.

public trustee An official appointed pursuant to the Public Trustee Act 1906.

puffing In auctions usually taken to mean exaggerating a property's good points.

puisne judge A judge of the High Court.

puisne mortgage A mortgage not protected by deposit of title deeds. It is a second or subsequent mortgage.

purchase The acquisition of land by a legal voluntary act such as conveyance as opposed to intestacy or bankruptcy.

purchase notice Land which has become incapable of reasonable beneficial use can in certain circumstances by required to be purchased by service of a notice on the appropriate local authority. See Town and Country Planning Act 1990 Part VI, as amended.

purchaser A person who acquires land by purchase. Under the Law of Property Act 1925 section 205(1)(xxi) the term purchaser includes a lessee and mortgagee.

purpose trust A trust that is neither for the benefit of a person or persons, nor is it a charitable trust. With limited exceptions these are void.

Q

qualified acceptance Acceptance of a bill of exchange that varies the effect of the bill as drawn.

qualified audit report Audit report in which some qualification of the financial statement is required.

qualified property Special property.

qualified valuer (1) A professionally qualified person who engages in the valuation of land or buildings. (2) Under the Royal Institution of Chartered Surveyors Guidance Notes for the Valuation of Assets, a qualified valuer is defined as 'a corporate member of the Royal Institution of Chartered Surveyors, or the Incorporated Society of Valuers and Auctioneers, or the Rating and Valuation Association, with appropriate post-qualification experience and with knowledge of valuing land in the location and of the category of the asset'.

qualifying costs In the context of housing associations, expenditure incurred by housing association in undertaking a development which contributes to the amount in which entitlement to Housing Association Grant is calculated.

quality control System imposed to ensure that each good or service reaches specified levels of quality.

Quango Acronym for quasi-autonomous non-government organisation. Quangos are semi-permanent public commissions or agencies who are usually answerable to a government minister.

quantity surveyor A qualified person who is responsible for producing a bill of quantities and dealing with all financial and contractual matters and is part of the design team and would have regular meetings with the contractor's quantity surveyor to discuss contractual issues and agree monthly and final accounts.

quantum 'An amount or the extent of'.

quantum meruit Literally, 'As much as is deserved'. A legal principle under which a person should not be obliged to pay, nor should another be allowed to receive, more than the value of the goods or services exchanged.

quarter landing A landing between floors extending the full width of one flight with a 90° angle.

quasi-easement A right which could amount to an easement if the dominant and servient lands were in separate ownership and occupation.

quasi-judicial Decisions made by administrative tribunals or government officials to which the rules of natural justice apply. These operate where a government policy-making body at times also exercises a licensing, certifying, approval or other adjudication authority, which is 'judicial', because it directly affects the legal rights of a person.

quasi-trustee A person answerable as trustee as she receives a benefit of a breach of trust.

que estate A dominant tenement.

Queen's Bench One of the three divisions of the High Court.

Queen's Counsel (QC) A senior barrister appointed by the Lord Chancellor.

quia timet Literally 'because he fears'. An injunction sought in anticipation of a threatened wrong. The applicant must show good cause for fearing the act will be committed against him or her.

quid pro quo Something for something. The giving of something in exchange for another thing of equal value.

quiet enjoyment A covenant for quiet enjoyment is implied into a conveyance for freehold land by virtue of Schedule 2 Part 1, Law of Property Act 1925. A covenant for quiet enjoyment is implied in every lease.

quinquennial Intervals of five years.

quit claim deed A deed or document operating as a release; intended to pass any title, interest or claim which the grantor may have in the property.

quo warranto Referring to a special legal procedure taken to stop a person or organisation from doing something for which it may not have the legal authority, by demanding to know by what right they exercise the controversial authority.

quorum The minimum number of members of an organisation who have to be present to make the body legally competent to transact business.

quoted price The official price for security or commodity.

R

rack-rent Open and full market annual value at the beginning of a lease.

rate A tax set by a local authority on the occupier of property in relation to its value. The rate levied on domestic properties was replaced by the community charge which has been replaced by the council tax.

rate demand The document issued by the rating authority to a ratepayer requiring payment of rates due.

rate of exchange The price of one currency in terms of another.

rate of return Annual amount of income from an investment expressed as a percentage of the original investment.

rate of turnover Accounting term for the speed with which stock is turned over. This is usually expressed in annual terms. The total sales revenue is often referred to as turnover.

rateable occupation Occupation which gives rise to a liability to pay rates.

rateable property Property on which the occupier is liable to be rated.

rateable value Defined under Section 19 of the General Rate Act 1967, as amended, as a figure upon which rate poundage is charged. It is (a) for those properties assessed direct to net annual value, an amount equal to the rent at which it is estimated, the hereditament might reasonably be expected to let from year to year that tenants undertook to pay all usual tenants rates and taxes and to bear the cost of the repairs and insurance and the other expenses, if any, necessary to maintain the hereditament in a state to command that rent or; (b) the gross value, less the statutory deductions for the purposes of the new local non-domestic rating which came into force on 1 April 1990, in England and Wales. Gross value becomes obsolete.

ratification Confirmation, for example of a contract, so as to make it binding.

rating agency Organisation which monitors the credit backing of bond issues and other forms of public borrowing.

Rating and Valuation Association An organisation established in 1882 whose membership consists of persons engaged in rating assessments, revenue collection, valuation and allied spheres of activity.

rating areas Section 1(1) of the General Rate Act 1967. They are defined as the boroughs and districts of every county; the City of London, the Inner Temple and the Middle Temple.

rating authorities Section 1(1) of the General Rate Act 1967 defines these as each borough or district council, the common council of the City of London, the sub-Treasurer of the Inner Temple and the under-Treasurer of the Middle Temple.

rating officer The official responsible for rating matters within a rating authority.

Rating Surveyors Association An organisation established in 1909 whose membership consists of surveyors specialising in rating.

rating year A period of 12 months beginning 1 April.

ratio analysis The use of ratios to evaluate a company's operating performance and financial stability.

ratio decidendi The relevant part of the judge's decision in a case, which is authoritative.

rationalisation Re-organisation of an organisation in order to increase its efficiency and profitability.

real estate Interests in land held by the deceased at the time of his or her death.

realised profit Profit which has arisen from a completed transaction.

realty or real property Land, including things attached to it. Realty includes immoveable property, such as a building or any object that, though at one time it may have been a chattel, has become permanently affixed to land or a building.

rebate Repayment of a sum of money because of default by the supplier.

rebuild The reconstruction of the entire, or part of a building, restoring it to its original form.

rebuilding clause The clause in a lease which requires the lessor or lessee to rebuild the demised properties in certain circumstances.

receiver A person appointed by a court or mortgagee under statutory powers, to safeguard property at risk and to collect the debts or rents due on behalf of mortgagees, debenture holders, or other creditors.

recitals Details of relevant earlier events attached to a deed which explain the background to transactions.

reconveyance Property subject to a mortgage prior to 1926 when it was conveyed to the mortgagee, who recovered the property on repayment of the loan.

record, conveyances by Conveyances of land effected by judicial or legislative act, as evidenced by the record, e.g. by Act of Parliament.

recourse The right of the holder of a note secured by a mortgage or deed of trust to look personally to the borrower or endorser for payment.

recovery High or county court procedure for the recovery of land from a person in wrongful possession.

recreational charity A trust for the public benefit providing facilities in the interests of social welfare. See Recreational Charities Act 1958.

rectification The correction of a document that does not accurately reflect the intention of the parties to it. Rectification can be made without application to court, if all parties agree and no third party's interests are compromised.

rectification of will The correction of the terms of a will which a court may authorise if satisfied that they do not fulfil the wishes of the testator, usually because of a misunderstanding or clerical error.

reddendum The clause within a lease specifying the rents to be paid, or the formula for its calculation.

redemption A right of redemption gives the vendor the right to buy back the property. On the other hand, where a mortgage is held, the release from the mortgage or completion of payments is known as redemption.

redemption date   Date at which certain category of security becomes due for repayment by the borrower. This is also referred to as the maturity date.

redemption period   The term during which a mortgagor may, by statute or agreement, redeem a mortgage.

redevelopment clause   A clause in a lease allowing for the redevelopment of property by one or other parties at a certain date.

re-entry   The lessor's right to reclaim the title to a property where there is a default in rental payments, or other breach of the lease agreement.

reference   The procedure where the Lands Tribunal is formally requested to determine a matter such as, firstly compensation, as a result of compulsory purchase of land, injurious affection, revocation orders, and secondly, valuations for taxation.

refinancing   The repayment of a debt from the proceeds of a new loan using the same property as security.

refurbishment   The modernisation, or improvement, of a building falling short of redevelopment or rebuilding.

regional planning guidance   Guidance from the Secretary of State for the Environment, Transport and Regions, which gives local authorities advice on the preparation of unitary development plans and structure plans.

register of charges   Registration with the Registrar of Companies of all charges on the property of a company by way of mortgages, fixed or floating debentures, bills of sale, etc.

register of companies   A register kept by Companies House of all companies who have filed documents with the Registrar of Companies and have obtained a certificate of incorporation.

register of directors   Register of directors giving names, former names, addresses, nationalities, business occupations, dates of birth and other directorships held. A copy of this register must be sent to the Registrar of Companies who must be informed of any alterations.

register of directors' interests   Register in which a company must detail the interests of its directors in the shares and debentures of the company.

register of interests in shares   Register which must be maintained by public companies showing interests in shares comprising 3% or more of any class of the voting share.

register of members   List of the members of a company which all UK companies are required to keep in their registered offices.

registered care home   This is a scheme which is registered under the Registered Homes Act 1984, and provides accommodation for elderly people, people with mental health problems, people with learning difficulties and those with a drug or alcohol abuse problem.

Registered Homes Act 1984   This Act requires any special needs housing projects with four or more residents which provide board and personal care to register with the local social services authority. The authority is required to inspect projects regularly and can insist on improved physical standards or staffing levels.

Registered Homes Amendment Act 1991   Act which incorporated small residential care homes within the auspices of the Registered Homes Act 1984.

Registered Housing Association   Housing association registered by the Housing Corporation under the provisions of Housing Association Act 1985, as amended.

registered proprietor   The person who is registered as the proprietor of registered land. (See registered title.)

registered social landlord   Term introduced by the Housing Act 1996 to apply to those housing organisations registered with the Housing Corporation. RSLs are required to register with the Housing Corporation if they wish to receive social housing grants. Additionally, to establish incorporation RSLs which are industrial and provident societies are registered with the Registry of Friendly Societies, charitable trusts are registered with the Charity Commissioners and charitable companies register with the Registrar of Companies.

registered title   A title registered under the Land Registration Act 1925. At the Land Registry the following titles must be recorded: fee simple, lease exceeding 21 years and an assignment of a lease with more than 21 years to run.

Registrar of Companies   Official charged with the duty of registering all companies within the UK.

Registrar of Friendly Societies   The registrar who has responsibility for enforcing the provisions of the Industrial and Provident Societies Act. Housing associations that are Industrial and Provident Societies must furnish an annual return to the registrar.

registration In the housing association context, housing associations have to register with at least two government agencies which have supervisory and regulated powers over them. The Registrar of Friendly Societies and Charity Commissioners and the Registrar of Companies, all must register with the Housing Corporation.

registration of a site In a housing association context, when an association identifies a site suitable for a housing project which it wishes to undertake, it registers it with the Housing Corporation, thereby preventing competition or duplication of effort among other associations

regulated tenancy A statutory or protected tenancy which is not a controlled tenancy. Residential tenancies created since 1989 are assured tenancies.

reinstatement The process by which a building is put back into the state it was at some relevant previous date.

reinstatement basis of insurance A method of loss assessment under the insurance policy where the amount payable is based on the cost of the insured property being repaired if damaged, or rebuilt if destroyed.

reinstatement value The value resulting from applying the reinstatement basis of insurance.

reinsurance The situation when an insurer insures the risk, or part thereof, with another insurer.

release An instrument releasing property from the lien of the mortgage, judgement, etc.

relief (1) A statutory reduction in liability to pay rates because of status of the occupier, or special circumstances relevant to the occupier's status, e.g. a dwelling altered to accommodate a disabled person. (2) Statutory reduction in liability for tax.

rem (See *in rem*.)

remainder An interest in land that comes into effect when a prior interests ends. A vested remainder is one in possession immediately, whereas a contingent remainder is one which is not determined until such time as the contingency is attained and the prior interest ends.

remainderman A person entitled to an estate in remainder.

remoteness Where an attempt is made to tie up property indefinitely beyond the period that the law allows. (See perpetuity.)

remuneration of trustees Not usually permitted except where authorised within the trust instrument or under statute (see Trustee Act s 41) or by the court.

renouncing probate The refusal of an executor to accept the role. (See probate.)

renovation Repair work carried out to an existing building to restore it to its original state.

rent This is usually a consideration paid to a landlord by a tenant in exchange for the exclusive use and enjoyment of land, a building or a part of a building. Under normal circumstances, the rent is paid in money and at regular intervals.

Rent Act 1977 This Act is relevant to the private sector and consolidates earlier legislation relating to security of tenure and rent control. For housing associations, its importance is in the parts that govern the operation of the fair-rent system and rent officer service. Fair rents are regulated under this Act, but since the Housing Act 1988, housing association and private levied tenancies are not subject to fair rents.

rent allowance In housing association context, the amount of money which an individual tenant obtains to help pay the rent to the housing association or private landlord. Since 1993 rent allowances have been included in the housing benefit.

Rent Assessment Committee A committee established to assess rents and agree terms of the statutory periodic tenancy under the Housing Act 1988 and to hear appeals from the decision of the Rent Officer under the Rent Act 1977, as amended.

rent assessment panel Persons appointed by the Secretary of State for the Environment and Lord Chancellor who are drawn on to be members of leasehold valuation tribunals, rent assessment committees and rent tribunals.

rent book The book which must be provided to record payments for every residential tenancy and which must contain certain specified information. Under sections 4–7 Landlord and Tenant Act 1985, as amended, it is a criminal offence not to provide such a book.

rent control The rent of private residential tenancies created before 1989 is restricted to a fair rent according to the Rent Act 1977 section 70. After that date under section 14 of the Housing Act 1988, as amended, the rent of such tenants including housing association tenants is restricted to market rent.

rent free period The period agreed during which a lessee is allowed occupation without payment of rent.

rent guarantee Guarantee given by a landlord concerning its intentions for future increases in rent, e.g. large-scale voluntary transfer housing associations often offer a five-year rent guarantee at transfer.

rent officer An official appointed by, but independent of, the local authority who, under the Rent Acts, determines and registers fair rents following an application by the landlord or the tenant, or both, under a regulated tenancy of a dwelling house.

rent pooling Pooling of resources to achieve a common purpose. This practice ensures that tenants can be charged similar rents for similar accommodation.

rent rebates Rent allowances granted to needy tenants under the Housing Finance Act 1972, the Furnished Lettings (Rent Allowance) Act 1973 and the Rent Act 1974. This scheme is supplemented by housing benefit under the Social Security Act 1986.

rent review The provision in a lease where the rent is reconsidered at stated intervals. The procedure for reviewing the rent is stated in the terms of the lease.

rent roll (1) The total amount of rent payable under the tenancies of one estate. (2) A method of recording information concerning the details of the rents payable from the tenancies of an estate.

rent to mortgage scheme Scheme which gives secure tenants of local authorities and non-charitable registered social landlords, who have a right to buy, a right to convert their rent into a mortgage payment so enabling them to buy a share in the property. This scheme was introduced by the Leasehold Reform Housing and Urban Development Act 1993.

rent tribunal A rent assessment committee sitting as a rent tribunal to determine rents for restricted contracts under the Rent Acts, as amended by the Housing Act 1980.

rentable area Floor space for which rent is calculated.

rental value The open market rent that might reasonably be expected to accrue subject to the terms of the lease.

repair The making good of defects in property which has deteriorated from its original state. Legally its meaning must fall short of effectively reconstructing or improving buildings.

repairs notice (1) A notice served by a local housing authority under the Housing Act 1985, as amended, requiring the execution of works to make a house fit for human habitation. (2) A notice served by the local planning authority requiring works to be carried out to a listed building for its proper preservation.

replacement cost The cost of replacing damaged items with new ones, often used in association with insurance contracts.

reply A pleading from a plaintiff in answer to the defendant's defence.

republication of will The re-execution, following the required formalities, of a will. The will can then be effective from the date of its republication and any alteration validated.

repudiation An express or implied refusal by one party to perform his obligation under a contract.

requisitions on title Questions raised by a purchaser of land concerning the vendor's title. This could include identity of property, outstanding mortgages, execution of deeds, etc.

rescision The termination or cancellation of a contract placing the parties to it in a position as if there had not been a contract.

resealed probate The granting of probate in one country, then approved in another, enabling an executor to deal with the testator's property in another.

reserve price The lowest price which a seller is willing to accept for an article offered for sale by public auction.

reserves The value of shareholders' funds in excess of the par value of their shares.

resettlement units Hostels run on a direct access basis and usually occupied by homeless single people in need of emergency housing.

residence The place where a person or persons live.

residence order An order under section 8 Childrens Act 1989 which settles the arrangements as to the person with whom a child is to live.

residential occupier An occupier of a residence under statute giving them the right to remain.

residual method of valuation A process of valuation of the property which has potential for development, or redevelopment. The estimated total cost of the work, taking account of fees and an allowance for interest, developer's profit and risk, is deducted from the gross value of the completed project. The resultant amount is then adjusted to the date of valuation, to produce the residual value.

residual value The value determined by the residual method of valuation.

residuary devisee The beneficiary entitled under a will to all real property not included in specific gifts in the will.

residuary estate Any part of a testator's estate that is not specifically bequeathed.

residuary legatee A person to whom the residue of a testator's personal property is left after the payment of debts, funeral expenses and legacies.

residue What remains of a deceased person's estate after payment of debts, funeral expenses, legacies and administration costs, etc.

respondent The defendant in certain types of action, including an arbitration.

restrictive covenants Private restrictions limiting the use of real property. Restrictive covenants are created by deed and may 'run with the land', binding all subsequent purchasers of the land, or may be 'personal' and be binding only between the original seller and buyer.

restrictive trade practices Agreements between traders which are not considered to be in the public interest.

resulting trust A trust that is presumed by the court or can arise automatically from certain situations. For example, a resulting trust is created when the beneficial interest is returned to the settlor as not having been fully disposed of. The court presumes an intention to create a trust; the law assumes that the right person does not hold the property and that the possessor is only holding the property 'in trust' for the rightful owner.

retail price index Index number which attempts to measure changes in the cost of living.

retained agent An agent retained by the principal with an instruction to sell, let or seek property. The payment of fees is normally on a contingency basis upon successful performance of the service.

retained profits The part of the net profit of a business or organisation not distributed to the shareholders but reinvested in the business.

retained sum The agreed percentage under a building contract which is to be retained by the employer for a specified period and paid only as and when specified defects have been satisfactorily remedied.

retaining wall A wall that retains and supports a mass of material on one side, such as soil.

return A change in direction of a wall at a corner and is normally set at 90°.

revaluation (1) A subsequent valuation. (2) The process of reassessing all hereditaments to a specified date for the purposes of rating, thereby creating a new valuation list.

revaluation of assets Revaluation of the assets of an organisation to reflect real and nominal changes in value.

revenue deficit grant A grant which is discretionary and payable to housing associations under the provisions of the Housing Act 1985 to cover the difference between rental income and outgoings on the managing and maintaining of properties and loan charges.

Revenue Support Grant Grant paid to local authorities by central government intended to ensure that similar standards of service are provided across local authorities.

reverse frontage A site or building's frontage to a side road or footway as distinguished from the main frontage, which is normally situated along the more prominent highway.

reverse gazumping The situation in which a purchaser breaks his word and fails to enter into a contract which fairly reflects the terms agreed but perhaps being willing to proceed with the original purchaser's price at a lower price. This situation is more likely to occur in a falling market when the advantage is given to the purchaser to acquire other property on better terms.

reversion A future interest left in a transferor or his (or her) heirs. A reservation in a real property conveyance that the property reverts to the original owner upon the occurrence of a certain event. It differs from a remainder in that a remainder takes effect by an act of the parties involved, whereas a reversion takes effect by operation of the law.

reversionary income Normally applies following a rent review or renewal of a lease where there is a potential change in income.

reversionary interest Any right in land which is deferred.

reversionary investment An investment property where the whole, or a substantial part of the capital value, is attributable to the prospect of a reversionary increase in rent.

revival of will The restoration of a previously revoked will by re-execution with appropriate formalities or by a later codicil.

revocation of probate Cancellation by the court of a grant of probate on the grounds that an executor is incapacitated and cannot act, or because probate was obtained fraudulently.

revocation of will Action by a testator to revoke a will. It can be done at any time before death and may be effected by destruction of the will, the writing of a new will or the addition of a codicil.

revolving credit Bank credit granted for a specified period within which draw down and repayment may take place. Repaid amounts may be redrawn up to the agreed credit limit.

rigging the market Action to create artificial market conditions which will benefit the operator.

right of entry The right of resuming or taking possession of land by entry in a peaceable manner. See Law of Property Act 1925.

right of way The right of passage over land owned by someone else. Rights of way are of different kinds and the purpose often specified, e.g. footpath, bridleway, etc. A private right of way is either an easement or customary right of way.

right to buy A right conferred upon certain tenants of non-charitable housing associations and local authorities to purchase a dwelling in which they are living, at a discount, after a minimum period of residence.

right to light (See light.)

right to manage Right of local authority tenants to set up an organisation to manage their own homes.

rights issue Issue of extra shares to existing shareholders to raise money for the company.

rights, perfect and imperfect (See perfect rights and imperfect rights.)

riparian Associated with a stream or river, e.g. land abutting a stream.

164

riparian owner An owner of land bordering on a watercourse.

riparian rights Rights pertaining to the banks of a river, stream, waterway, etc. The rights of persons who own land that runs into a riverbank. While not a true ownership right, riparian rights include the right of access to and use of the water for domestic purposes. The extent of these rights varies and may include the right to build a wharf outwards to a navigable depth or to take emergency measures to prevent flooding.

rising rent A rent which increases upwards only by pre-determined amounts at stated times during the terms of the lease.

risk The items in an insurance agreement against which protection is provided.

risk capital Also known as venture capital. This refers to capital invested in a project which contains an element of risk.

roll over relief The postponement of Capital Gains Tax payments where the proceeds, or part of the proceeds, of disposal of business assets, are used to acquire other assets.

rolled up interest The situation where interest is not paid at intervals, but instead is added to the principal amount of a loan as it accrues. Often used in association with a development project allowing for payment on completion of the project after sales of the building, or buildings.

rolling budget Also known as a continuous budget. This refers to a budget which is constantly updated in which as one period ends another is added, e.g. a one-year budget would have a new month or quarter added as each month or quarter ends.

rotation of directors Under the Articles of Association it is normal practice for a number of directors to retire each year and offer themselves for re-election. Shareholders therefore have an opportunity to change directors. It is usual for one third of directors to retire each year.

Royal Institute of British Architects The professional body established in 1834 and granted a Royal Charter in 1837, its purpose being the advancement of civil architecture and facilitating the acquiring of knowledge connected therewith.

Royal Institution of Chartered Surveyors The professional body established in 1868 concerned with surveying. It was granted a Royal Charter in 1881. The Institution is divided into various faculties, these being building surveyors, general practice surveyors, land agency surveyors, agriculture surveyors, land surveyors, minerals surveyors, planning and development surveyors, and quantity surveyors.

Royal Town Planning Institute   The professional body in the United Kingdom established in 1914 concerned with town planning.

RSL   Abbreviation for Registered Social Landlord.

rule in Allhusen v Whittell   Presumption that where a testator settles his residuary estate for persons in succession, she intends each of these persons successively to enjoy the same property, i.e. the net residue estate after payment of debts and funeral/testamentary payments.

rule in Cradock v Piper   Under this rule, a solicitor who is a personal representative is entitled to profit costs for litigation work done on behalf of the estate, except so far as these costs have been increased by the solicitor being one of the parties.

rule in Lassence v Tierney   Where there is an absolute gift to a legatee onto which trusts are engrafted or imposed which fail for any reasons, then the absolute gift takes effect.

rule in Strong v Bird   Where the deceased had a continuing lifetime intention to give specific property to another but failed to complete an effective lifetime transfer the gift will become perfected if the donee of the gift is appointed as the deceased's executor or personal representative (Re James).

running with land   (See covenant.)

Ryde's scale   The scale of fees paid to valuers for work carried out in preparing claims for compensation and negotiation of their settlement following the compulsory acquisition of land.

S

salary   Regular payment made by an employer under a contract of employment to an employee.

sale   Transfer of property in consideration for a sum of money.

sale and leaseback Arrangement whereby a company or other organisation sells an asset and immediately purchases from the buyer a right to use the asset under a lease agreement. This arrangement is often entered into when a business needs the cash tied up in its lands and buildings.

sale price The amount realised on the disposal of the property.

sale, power of The right to sell the property of another, usually in association with settlement of a debt that is owed and via a statutory procedure, e.g. charging order.

sales agreement See agreement of sale.

sales forecast Estimate of future sales volume and revenue.

sales journal In book keeping the book of prime entry in which an organisation records invoices issued to customers for goods or services supplied.

sales ledger Ledger which records the personal accounts of an organisation's customers.

sales volume profit variance The difference between the actual units sold and those budgeted, valued at the standard profit margin.

salvage of trust property Where absolutely necessary the court may permit the sale or mortgage of part of a minor's beneficial interests for the benefit of the property to be retained.

satisfaction (1) The fulfilling of a claim. (2) The doctrine that payment, performance or some other act discharges an obligation.

satisfaction of mortgage Document issued by mortgagee when the mortgage is paid off.

Saunders v Vautier (1841) Where together the beneficiaries are *sui juris* and absolutely entitled to the trust property they can terminate the trust and call for the trust property to be transferred to themselves or to a third party.

save as you earn (SAYE) Method of making regular savings from pay which carries tax privileges.

scale Units used on a drawing to represent the actual size of a building element, e.g. a scale of 1:5 means that each unit drawn is five times smaller than the actual size.

scale fees The charges authorised as appropriate by a professional body for specific types of work. It is no longer possible for such scales to be accepted as minimum fee scales, except in limited cases.

scale rule A special rule containing a number of scales for producing or reading dimensions from drawings. (See scale.)

schedule of condition A document indicating the physical state of a building often used in association with the assessment of compensation under a compulsory purchase procedure.

schedule of dilapidations A document listing the requirements of repair and maintenance which a tenant or landlord is obliged to make good under the terms of a lease or tenancy.

schedule of prices A priced specification of works.

schedule of works (See specification of works.)

scheduled monument A monument registered in a schedule of monuments maintained by the appropriate Secretary of State under the Ancient Monuments and Archaeological Areas Act 1979, as amended.

scheme audit In a housing association context, an analysis by the Housing Corporation of new housing association schemes on a sample basis after their completion to determine whether an association has performed its procurement function correctly.

scheme performance criteria In a housing association context, a set of standards to which new publicly funded housing association schemes must conform.

Scottish Homes Organisation formed by the merger between the Scottish Special Housing Association and the Scottish Office of the Housing Corporation. It assumes the functions and powers of the Housing Corporation in Scotland under the Housing (Scotland) Act 1988 and manages properties formerly owned by Scottish Special Housing Association.

scrip issue The re-division of a company's capital into more manageable units and their issue to the company's shareholders in proportion to existing holdings.

seal An impressed piece of wax attached to a document as to make it under seal.

sealing Used in the execution of documents based on indicating assent.

searches Examination of registers and records to determine encumbrances affecting the title to property. During the conveyance of a title in registered land a search is made at HM Land Registry. For unregistered land a search is made at the Land Charges Department. Local land charges searches are made for both registered and unregistered land. It is good practice, depending on the circumstances to make other searches, such as company search, index map search, coal mining search, commons registration search, etc.

second home Commonly known as a holiday home, this home is not rented and is occupied occasionally by the owners.

secondary co-operative In a housing association context, an organisation, normally a registered housing association, supplying development, financial and other services to primary co-operatives.

secondary financing A loan secured by a mortgage or trusts deed, which lien is junior to another mortgage or trust deed.

secret profits Profits made by an agent which are not accounted to her principal.

secret trust A trust, the existence and terms of which are unknown, whereby a testator gives property to a person on that person's promise to hold it on trust for a third party.

section A drawing to scale which shows the construction of the various parts of a building from the foundations to the roof.

secure tenancy A tenancy where the tenant has security of tenure by virtue of the Housing Act 1985, as amended.

secured loan Monies advanced by way of a loan but where on default by the borrower, the lender has a legal right to sell property against which the loan has been secured.

Securities and Futures Authority Limited (SFA) Self-regulating organisation responsible for the conduct of brokers and dealers in securities, options and futures.

Securities and Investment Board (SIB) Regulatory body set up by the Financial Services Act 1986 with responsibility for the overall regulation of the investment industry.

security Real or personal property pledged by a borrower as additional protection for the lender's interest.

security of tenure A statutory right of a tenant to continue occupancy after the completion of a term has expired.

seed capital Minimum amount of initial capital required to fund the research required before a new company is set up.

self-assessment System which enables individuals to assess their own liability for income tax and capital gains tax.

self-build society The Housing Associations Act 1985 defines it as 'a housing association whose objective is to provide for sale to, or occupation by, its members, dwellings built or improved principally with the use of its members' own labour'.

self-dealing Used to describe the circumstances where a trustee acting for herself but in her capacity as trustee places herself in a position where her obligations conflict with her self-interest.

self-employed Individuals who are not employees and who trade on their own account.

seller's broker Agent who takes the seller as a client, is legally obligated to a set of fiduciary duties and is required to put the seller's interests above all others.

sellers market Market in which demand exceeds supply, a situation which enables sellers to increase prices.

sellers over Stock exchange term to describe market condition where the number of sellers exceeds buyers.

selling price The amount which a vendor states that he is willing to accept whether or not it is the price eventually realised.

selling price variance The difference between the actual selling price per unit and the standard selling price per unit multiplied by the actual number of units sold.

semi-detached house Two separate dwellings which are joined by a separating or party wall.

semi-variable costs Costs which include both fixed costs and variable cost components, i.e. costs which vary to some extent with the level of activity but not in direct proportion, e.g. the consumption of electricity includes an element of both fixed and variable costs.

sensitivity analysis A technique used to analyse the sensitivity of a decision to a change in one or more of the assumptions used in making that decision.

separating wall (See party wall.)

separation agreement An agreement concerning the disposition of a matrimonial home.

septic tank An underground tank in which sewage from the house is reduced to liquid by bacterial action and drained off.

sequestration Order of the High Court to seize goods and lands of the defendant who is in contempt of court.

service charge The sum payable by a tenant for services provided by the landlord.

service contract agreement In the housing association context, a legal contractual agreement using a partnership scheme to provide special needs housing and defines the responsibilities of the Housing Association as a voluntary agency.

service occupancy An occupancy, normally of residential premises, which is granted by an employer to an employee as part of his conditions of employment for the better execution of his duties which continues only during the period of employment.

service tenancy A tenancy where the tenant is an employee of the landlord. The employment being the motive for the grant of a tenancy.

services Gas, water, electricity, drains and other provisions serving a building.

servient tenement A tenement subject to an easement.

servitude An easement.

set-off Diminution or extinction of the plaintiff's claim in an action by deducting the defendant's counter claim.

setting aside A court order that cancels or makes void another order or judgement.

settled land Land limited to certain persons in succession. Under section 1 Settled Land Act 1925 it is defined as: (i) Limited in trust for any person by way of succession. (ii) Limited in trust for any person in possession, (a) for an entitled interest, (b) for an estate subject to an executory limitation over, (c) for a base or determinable fee, (d) being a minor. (iii) Limited in trust for a contingent estate. (iv) limited to or in trust for a married woman with a restraint on anticipation. (v) charged with a rent charge for the life of any person.

settlement A disposition of land made by deed or will. Trusts are created and designate the beneficiaries and the terms under which they are to take the property. Previously they could exist as strict settlements under the Settled Land Act 1925 or trusts for sale under the Law of Property Act 1925. The Trusts of Land and Appointment of Trustees Act 1996 now prohibits the creation of new strict settlements, but those which exist at the date when the Act came into force remain valid. All settlements created after 1 January 1997 will take effect as trusts of land.

settlement, accumulation and maintenance (See accumulation and maintenance settlement.)

settlor A person who makes a settlement in regard to property.

several Where two or more persons share an obligation so that it may be enforced in full against any one of them, independently.

sex discrimination The singling out of a person or group because of gender for special favour or disfavour. Discrimination may be direct, indirect or by victimisation. It may be unlawful under the Sex Discrimination Act 1975.

sexual harassment (See harassment.)

share capital Capital of a company which arises from the issue of shares.

share certificate Document which provides evidence of ownership of shares in a company.

share premium An amount over and above the nominal value of a share charged at issue. Share premiums are credited to a special share premium account which is the capital reserve account.

share register The register kept by a limited company which details names and addresses of members and the extent of their shareholding.

shared housing Residential accommodation in which two or more individuals live together and share facilities. The term also includes hostels.

shared ownership In a housing association context, the sharing of the equity in a property between the occupier and housing association. The occupier purchases a property at a proportion of the value and pays rent to cover the share in the equity retained by the housing association. Typically the purchaser acquires 50% equity and pays

50% at normal rent. When the purchaser leaves he realises the capital invested.

shell company Company which exists only in name.

Shelter Established in 1965 following public concern about the plight of homeless people. It is responsible for campaigning on behalf of the nation's homeless and raises money from various sources to support its campaigning roles, and to finance a national network of housing aid centres and to support individual projects aimed at alleviating homelessness.

sheltered housing A colloquial term used to mean housing specially designed for elderly people grouped around a range of communal facilities. (See Category 1 housing and Category 2 housing.)

sheriff A local office of great antiquity. The present day duties of the sheriff in civil cases is the execution of judgement.

shifting use A use which shortens a preceding interest.

short bill A bill of exchange payable on demand at sight or within 10 days.

short lease For income tax and capital gains tax purposes a lease for a term of 50 years or less.

short life housing Properties which can be used temporarily, for example, student accommodation or single accommodation.

short tenancy In relation to compulsory purchase a tenancy for not more than a year.

short term interest rates Rate of interest payable on short term loans.

shorthold tenancy A tenancy of a residence under which the landlord has the mandatory ground for obtaining possession that it is a shorthold tenancy. (See protected shorthold tenancy.)

SI units The international system of metric units used for scientific and technical purposes.

sight line A line used on a plan to establish visibility standards at road junctions or access points on to public roads. The basic criteria is that at eye level, being 1.05 metres above road level, there should be a clear view over a given area.

signature A person's name, sign or other mark impressed on a document indicating an intention to be bound by the document.

signing dates The date on which a valuation report or certificate is signed.

simple contract A contract not under seal, whether oral or in writing.

simple trust (See bare trust.)

site An area of land to be used for a building project.

site coverage The proportion of a site which is covered by buildings, often expressed as a percentage of the area covered by buildings to the total area of a site.

site investigation A survey carried out to gather facts and information of the building site used for design purposes and, in particular, foundation design.

site plan A drawing in the horizontal plane of an area of land showing boundaries and physical extent of the land included in a particular parcel.

site value The value of undeveloped property as a site for future development.

sitting tenant A tenant who is lawfully in physical possession, or is entitled to immediate physical possession of the demised premises.

sitting tenant value The price which might reasonably be expected to be secured on the open market at a given time by the tenant currently in possession of the land.

slum clearance The procedure for designating or acquiring and demolishing unfit buildings.

small company A company which meets the criteria for small companies contained in the 1989 Companies Act. Such companies enjoy advantages in relation to the statutory requirements regarding the filing of accounts and other documents with the Registrar of Companies.

SMM (See standard method of measurement.)

social cost Those costs of a product which are not directly paid by its producer but by society in general, e.g. environmental costs.

social housing standard Housing Corporation standards and requirements for the production of new social housing. These include rents, maintenance, tenants' rights, allocations and development standards.

soft currency Currency for which there is relatively little demand usually because of a weak balance of payments situation.

sole agent A person who is the only agent entitled to represent his principal.

sole selling/letting right The right affected by an agent when the principal has contracted to convey exclusive rights to sell or let the property, entitling the agent to a commission even if the principal acts on his own behalf.

solicitor A legal practitioner, one who offers legal advice. For a person to practise as a solicitor, they must: (1) Be admitted as a solicitor. (2) Have his or her name on the roll of solicitors. (3) Have in force a practising certificate issued by the Law Society.

solvency State in which an organisation is able to pay its debts as they fall due.

special administration The administration of specific effects of a deceased person, sometimes called limited administration.

special damages Ascertained or calculated monetary loss, as opposed to unascertained or general damages. Damage of a kind that must be expressly pleaded and proved before the court.

special lien A lien that binds a specified piece of property, unlike a general lien, which is levied against all one's assets. It creates a right to retain something of value belonging to another person as compensation for labour, material, or money expended on that person's behalf.

special needs housing Residential accommodation provided for people with a special disability or requirement, in addition to their need for a home. It is either self-contained accommodation or shared housing.

special needs management allowance In a housing association context, an allowance paid to registered housing associations to cover the additional cost of special needs schemes established since the Housing Act 1988.

special personal representative A personal representative appointed to act in respect of settled land only.

special projects promotional allowance In a housing association context, a Special Housing Association Grant Allowance to cover some of the additional costs incurred when associations develop special needs schemes.

special trust A trust imposing active duties on the trustees.

special warranty deed A deed in which the grantor conveys title to the grantee and agrees to protect the grantee against title defects or claims asserted by the grantor and those persons whose right to assert a claim against the title arose during the period the grantor held title to the property. In a special warranty deed the grantor guarantees to the grantee that he has done nothing during the time he held title to the property which has, or which might in the future, impair the grantee's title.

specific devise (See devise.)

specific legacy (See legacy.)

specific performance A court order that an individual must follow the terms of a contract to which she is a party.

specification A description contained in a specification of works.

specification of works Part of the design information and describes the work involved in carrying out the various parts of the building work and is either a separate document or used to supplement a bill of quantities.

speculative development The development of the building or buildings where at the start of the project there is not a known buyer or tenant.

speculative funding The funding of a future development where there is a risk of not selling or letting at a figure exceeding, or even attaining, the development expenditure.

speculator A person who undertakes transactions in property with a view to making a profit but with the risk of making a loss.

spot listing The emergency process whereby a building is added to the list of buildings of architectural and/or historic interest. The process is usually initiated by the local planning authority in response to a real or perceived threat to an appropriate class of building.

spouse A husband or wife.

squatter Someone who occupies land and is not a tenant, licensee or otherwise without permission from the person who is entitled to occupation.

176

squatters title The title to land acquired by 12 years adverse possession against the person who has lawful possession of the land.

stag Stock exchange term for individual who buys a new issue of stocks and shares in expectation that the price will rise very quickly so producing a profit.

stale cheque A cheque which has not been presented for payment within six months of issue which the bank will not honour.

stamp duties Duties raised by affixing stamps to instruments such as conveyances and leases.

standard cost Predetermined cost that management establishes and uses as a basis for comparison with actual costs.

standard deviation A measure of the dispersion of statistical data which gives proportional significance to extreme values in the series being examined.

standard method of measurement A document containing the nationally recognised method for measuring quantities from drawings and is used to produce a bill of quantities.

stare decisis Literally 'to stand by things decided'.

starter homes Residential accommodation specifically designed and constructed for first-time buyers.

State Earnings Related Pension Scheme (SERPS) Government run scheme which aims to provide an earnings related pension for every employed person. Individuals may contract out of SERPS and subscribe to an occupational pension scheme or personal pension scheme.

statement of account Record of the transactions of an organisation with a customer for a specified period which normally shows the indebtedness of one to the other at the end of that period.

Statement of Accounting Standards (SAS) Statement issued by the Auditing Practices Board on the principles and procedures in auditing.

Statement of Standard Accounting Practice (SSAP) Accounting standards prepared by the Accounting Standards Boards and issued by the members of the consultative committee of accountancy bodies.

statute An Act of Parliament.

Statute of Frauds The law requires that certain contracts, such as agreements of sale, be in writing in order to be enforceable.

statutory books The books which a company is bound to keep as a requirement of the Companies Act 1985. These include Register of Members, Register of Directors, Register of Charges and Minute Books.

statutory declaration A declaration made before a person authorised to administer an oath but not made in court.

statutory form of accounts In a housing association context, the formula in which the accounts of registered housing associations must be published as set out by statutory instrument under the Housing Associations Act 1985.

statutory instrument A form of delegated legislation which has the full force of the law.

statutory lien An involuntary lien includes tax liens, judgement liens, mechanic liens, etc.

statutory lives in being Lives set out in the Perpetuities and Accumulations Act 1964, s 3(5) for the purposes of the perpetuities rule.

statutory sick pay A compulsory scheme operated by employers under which payments are made to individuals absent from work as a result of sickness.

statutory tenant A tenant who remains in possession of a residence after the proper term has been completed by virtue of the Rent Act 1977, as amended.

statutory trusts Trusts created or implied by statute, e.g. under the Administration of Estates Act 1925.

statutory will A will ordered to be executed by the Court of Protection for an adult patient who the court has reason to believe is incapable of making a valid will for herself.

staying put Colloquial term used to describe agency services which provide assistance to elderly home owners with housing repairs, improvements and adaptations.

stepped costs Fixed costs which are unchanged within certain limits of activity but step up or down to a new level when these limits are exceeded.

stirpital substitution A clause in a will providing that if a child does not survive to take her share, her issue will take in her place.

Stock Exchange See London Stock Exchange.

stock turnover Ratio which measures the speed at which raw materials or finished goods are being consumed or sold. This is calculated by dividing the total sales for a stated period by the total average stock carried during that period.

stop notice The part of the planning enforcement process defined under the Town and Country Planning Act 1990 as: 'Where the local planning authority consider it expedient that any relevant activity should cease before the expiry of the compliance with an enforcement notice, they may, when they serve a copy of the enforcement notice or afterwards, serve a stop notice prohibiting the carrying out of that activity on the land to which the enforcement notice relates, or any part of that land specified in the stop notice.'

straight line method Method of calculating depreciation in which a fixed asset is depreciated by an equal amount each year over its expected life.

street A term generally meaning a made-up road which has houses, or other buildings, on one or both sides. The term, however, has different meanings in different statutory circumstances.

structural design The process involved in the designing of the structural parts of the building such as load-bearing walls and beams.

structural engineer A qualified person who is responsible for structural design.

structure plan A strategic plan for the future development of an area, usually prepared by a county planning authority. The document does not contain site specific landuse policies.

sub lease Any lease granted below the head lease. The person with the benefit of the sub lease is the sub-tenant. The sub lease cannot be for a period longer than the lease since the tenant has no power to grant a lease which is longer than his own interest in the property.

subpoena An order requiring a person to appear in court and give evidence or produce documents.

subrogation The right to bring an action in the name of another person.

subscribing witness A person who signs a document as an attesting witness.

subsequent condition (See condition subsequent.)

substitute of trustee A document which is recorded to change the trustee under the deed of trust.

substitutional legacy A legacy that passes gifts to the descendants of a beneficiary where the beneficiary predeceases the testator.

sub-trust Where the absolute beneficial owners declare a trust of their beneficial entitlement.

succession Where a person becomes entitled to property previously enjoyed by another.

sue To take legal proceedings for a civil remedy.

sui juris Literally 'of his own right', the term used to describe a person who has full legal capacity.

sum insured The maximum sum of an insurer's liability in terms of an insurance contract, being determined by adjustment or settled on request of the insured.

summons An order to appear before a judge or magistrate. Some civil actions are begun by an originating summons.

supervision order An order whereby a minor is placed under the supervision of a local authority or probation office. See Children and Young Persons Act 1969 sections 7 and 11 as amended by the Childrens Act 1989, Criminal Law Act 1977 section 65(4), Powers of Criminal Courts Act 1973 section 26 and Matrimonial Causes Act 1973 section 44, as amended.

supreme court The High Court, the Court of Appeal and the Crown Court are collectively known as the Supreme Court of Judicature.

surety bond In a housing association context, a guarantee purchased by a housing association to provide cover if a contractor is unable to complete a contract.

surrender The yielding of a lease of land. Normally executed by deed.

surrender clause The clause in a lease obliging the tenant to surrender or offer to surrender the lease.

surrender value Value of the unexpired portion of a lease if surrendered to the immediate landlord.

survey (1) A systematic investigation of a problem via measurement and/or assessment. (2) Map or plan made by a surveyor who measures land and charts its boundaries, improvements, and relationship to the property surrounding it. A survey is often required by the lender to assure him that a building is actually sited on the land according to its legal description.

surveyor A qualified person who would normally be a member of the Royal Institution of Chartered Surveyors and would be involved in one of the surveying disciplines. (See quantity surveyor, building surveyor, land surveyor.)

survivors In relation to succession construed as meaning those who are living at the period of distribution.

survivorship, right of The right of a survivor to the whole property, e.g. where property is held by joint tenants.

system building Building methods adopted in the 1960s which consisted of prefabricated components manufactured in a factory and delivered to site for quick assembly. This method was generally used for building multi-storey blocks of flats for social purposes and many systems were developed between consortiums of building organisations and local authorities to meet social housing demands much more quickly than could be achieved by traditional methods. Due to ensuing social problems, poor design and high maintenance costs, system building is not now used and many system-built flats are being demolished to make way for traditional housing.

T

tacking The priority of mortgages over the same property being determined by order in which the mortgages were made. Repealed by the Law of Property Act 1925.

Tai Cymru Organisation established to take over the function of the Housing Corporation in Wales which is also known as Housing for Wales. Its duties are set out in the Housing Associations Act 1985 and amended by the Housing Act 1988 and Housing Act 1996.

takeover bid Offer made to the shareholders of a company by a person or organisation to buy their shares at a specified price in order to gain control of the company.

taking off The measurement and listing of quantities of materials from drawings for the purpose of preparing a bill of quantities in accordance with the Standard Method of Measurement of Building Works (SMM), or for estimating, planning and ordering purposes.

tangible assets Physical assets in the ownership of the individual or business. Whilst this term strictly applies to those assets which can be touched it is often used to describe fixed rather than current assets.

tangible property Property which can be touched, that is, it has a physical existence, as compared with intangible property, e.g. choses in action, patents, etc.

taper Adjustment made to benefits according to the income of a claimant.

tax avoidance Action to minimise total taxation liabilities which is legal.

tax evasion Action taken to minimise tax liabilities illegally by not disclosing full information to the relevant authorities.

tax point Point in the distribution of goods and services at which Value Added Tax becomes payable.

tax return Form upon which a taxpayer makes an annual statement of income and personal circumstances enabling a calculation of tax liability to be made.

tax tables Tables issued by the Inland Revenue to assist employers in calculating the tax due from employers under the pay as you earn system.

taxable income Income liable to taxation.

taxation of costs The process where the disputed costs of a court action, public inquiry, etc. are reviewed by the relevant officer of the court to determine the charges recoverable by one party from the other.

telegraphic transfer Method of transferring monies abroad by means of cabled transfer between banks.

tenancy A holding, as of land, by any kind of title, occupancy of land, a house or the like under a lease or on payment of rent or tenure.

tenancy at sufferance A tenancy which arises when a tenant holds over after expiration of his lease.

tenancy at will A tenancy which may be terminated at the will of either the lessor or lessee.

tenancy in common Where two or more persons have concurrent ownership, each having a distinct but 'undivided' share in property. No one person is entitled to exclusive use or title, each being entitled to occupy the whole in common with others. Since the Law of Property Act 1925 tenancy in common only exists as an equitable interest under a trust.

tenant Any person in possession of real property with the permission of the owner.

tenant at sufferance A person who has originally come into possession of land by a lawful title and continues possession after his interest has determined.

tenant at will A state of being between tenant and lessee.

tenant for life A tenancy which expires upon the tenant's death. Since 1925 a life tenant can only exist in equity.

tenant for years A lessee with a term of years certain.

tenant from year to year A tenant whose tenancy can only be terminated by a notice to quit on its anniversary. A six months' notice to quit is usually required.

tenant in the praecipe A tenant against whom a writ was issued in a real action. (See recovery.)

tenant-right The right of a tenant after determination of his tenancy.

tenants' association A voluntary association of tenants who live in a particular area or scheme.

Tenants' Charter A charter under the Housing Acts which gave tenants new rights which are security to sub-let and undertake improvements.

tenants' choice A colloquial term used to mean the process introduced in the Housing Act 1988 which gives secure local authority tenants in England and Wales the opportunity to change landlord while staying in their present homes.

tenants' guarantee Standards of service which social landlords are required by the Housing Corporation to provide to all types of tenants other than secure tenants.

tenant's improvements Improvements to land or buildings carried out wholly or partly at the expense of and to meet the needs of the tenant.

Tenants' Incentive Scheme (TIC) Cash payments made so that tenants of participating registered social landlords can buy a property on the open market on condition that the association offers the subsequent vacancy to a homeless family.

tenant's repairing lease A lease where the lessee is under an obligation to maintain and repair the property.

tenants' right to exchange The right of secure and assured tenants of housing associations to exchange their home with a similar tenant.

tenants' right to repair A legal right which allows secure tenants to claim reimbursement of the cost of urgent repairs which the landlord has failed to carry out.

tender (1) An offer of land, goods or services which if accepted creates a legally enforceable contract. (2) An offer made by a contractor and requested by an architect to carry out the building works in accordance with the design information for a certain sum of money.

tendering The procedure involved by a contractor in preparing a tender working from the design information.

tendering process The process of inviting bids by tender, receiving, considering and usually accepting one. The normal process is: (1) Preparation of identical tender documents stating details about the subject of the bid. (2) Dispatch of the tender documents to the tenderers, who may be selected or any one who requests and may have to meet certain criterion. (3) The contents of each bid are confidential until a time and date specified at the outset, when tenders are opened. (4) No tender is accepted after the date for opening and considering tenders.

tenement A thing which is the subject of tenure, e.g. land. A house let as different apartments.

tenure The type of holding of land, as in freehold tenure.

tenure, security of The right of a tenant or licensee of land to continue possession under a statute after determination.

term of years An interest in land for a fixed number of years.

terminal value Value of an investment calculated at the end of an investment period.

terra Land.

terraced house A dwelling within the same structure composed of more than two units, all having separate access at ground floor level.

testable A person legally capable of making a will or being a witness.

testament A will that disposes of a person's personal property but not land, usually refers to a will without that distinction.

testamentary capacity In law, the ability to make a will.

testamentary expenses Expenditure on the performance of the executor's duties.

testamentary freedom A person's right to dispose of his or her property as she or he wishes. See also the Inheritance Act 1975 which limits that freedom to some extent.

testamentary guardian A guardian appointed by will.

testamentary intention A valid intention to create a will which is essential for the validity of a will.

testamentary trust An express trust intended to operate after death.

testate On death, having left a valid will. (See intestacy.)

testator A deceased person who has made a will.

testatrix A woman who makes a will.

testatum The part of the deed where the witnessing takes place.

testimonium clause The final clause in a will or deed.

The Housing Finance Corporation (THFC) Independent intermediary body set up by the NHF and the Housing Corporation in 1987 to raise private finance for housing associations.

thing in action (See chose.)

third party A person not originally a party to an action, but who may be brought in by the defendant.

third surveyor A surveyor appointed to settle a party wall dispute, which has already by investigated by two surveyors, each representing separate parties to the dispute.

tied cottage A dwelling house provided to the occupant as part of, and effective only during, his employment. Usually associated with agricultural employment.

time is of the essence Legal phrase in a contract requiring punctual performance of all obligations.

Time-Share Developers Association An association of developers of time-share property.

timeshare accommodation Living accommodation used for leisure purposes by a group of people all of whom have rights to its use at intermittent times.

timeshare rights Rights accrued by a person who is a timeshare user, being rights exercisable during a period of not less than three years.

title Term signifying a right to property. The classes of title are: absolute title, which means in effect that it is as perfect as it can be; possessor title is the same as absolute title, except that the proprietor is also subject to all adverse interests existing at the time of first registration; qualified title is granted where the title submitted for registration shows a specific identified defect which the registrar deems to be of such a nature that he cannot use his discretion to overlook the defect and grant absolute title; good leasehold title applies to leasehold estates and is awarded where the registrar is satisfied that the title to the leasehold interest is sound, but having no access to the title to the superior reversionary interest, he is not prepared to guarantee the lease against defects in the freehold title, or to guarantee that the freeholder has the right to grant the lease.

title guarantee Full title guarantee is a covenant given on the disposition of land that the seller has the right to dispose of the land as he purports to and that he will do all he can to transfer title he purports to give. It also means that the land is disposed of free from encumbrances other than those the seller does not know about. Limited title guarantee, is the same as full title guarantee, except that the seller covenants that he has not himself encumbered the land, and also that he is not aware that anyone else has done so since the last disposition for value.

title search A review of the records, to make sure the buyer is purchasing a house from the legal owner and there are no liens, or other claims or outstanding restrictive covenants, which would adversely affect the marketability or value of title.

title-deed The documents proving title to land.

topography The natural features of the surface of a building site such as slopes and depressions.

topping out The completion of a building project to the point when the structure and frame, including the roof are completed. It is usually marked by a celebration and attaching a small tree to the highest point of the building.

tort A civil wrong, e.g. nuisance, negligence, trespass, etc.

total cost indicators In the housing association context, it is the value for money criteria used by the Housing Corporation for schemes developed under its current funding regime.

total profits Profits chargeable to corporation tax.

tower block A tall building, normally flats or offices which are eight storeys or more in height.

town and country planning The statutory process of planning and controlling the future development of an area introduced under the Town and Country Planning Act 1947, as amended.

Town and Country Planning Act 1947 This legislation brought almost all development under control by making it subject to planning permission. Planning went from being a 'regulatory function' and development plans were prepared for every area of the country.

Town and Country Planning Act 1990 This Act consolidated all previous planning legislation.

town house (See terraced house.)

town planner A person who engages in the profession of Town and Country Planning. Usually qualified by Corporate Membership of the Royal Town Planning Institute.

tracing trust property The process used by beneficiaries to recover trust property that has come into the hands of others.

trade A skilled occupation such as a joiner and bricklayer.

Trades Union Congress (TUC) Central UK organisation to which most trade unions are affiliated.

traffic generation The build-up or potential build-up of vehicular traffic resulting from an existing or potential land use.

training agency Organisation set up by the government in 1989 with responsibility for supervising the Training and Enterprise Councils (TECs) which are independent companies set up to actively promote training by employers and individuals with the aim of regenerating local economies. The agency is responsible to the Secretary of State for Employment and is government funded.

Training and Enterprise Councils (TECs) (See training agency.)

tranche An instalment or part of a total sum of money paid or advanced. A term used in association with development funding with each tranche of money paid at various stages of development in accordance with the terms of a contract and on the issue of certificates from an architect or surveyor.

TransAction The national protocol introduced by the Law Society in 1990 which standardised the procedures relating to domestic conveyancing.

transfer of engagements The situation where housing associations and the Industrial Provident Society transfer their engagements to another association which undertakes to fulfil these. No conveyance or assignment is needed and no stamp duty is payable.

transfer pricing Price paid for the internal transfer of goods or services between difference sectors of a multi-national corporation, individual subsidiaries within a group of companies or between the profit centres or trading units of a firm.

treasury bill Bill of exchange issued by the Bank of England on the authority of the government which is repayable in three months. Treasury bills are issued by tender each week to the discount houses.

tree preservation order An order made by the local planning authority under the Town and Country Planning Act 1990, as amended, to preserve individual or groups of trees. Such an order prohibits the cutting down, lopping, topping, uprooting or wilful destruction of the tree without the consent of the local authority.

trespasser One who goes on land without permission and whose presence is unknown to the proprietor, or if known is practically objected to.

trial balance List of the debit and credit balances on all accounts of an organisation at a stated date. The total of debit balances should equal the total of credit balances within the double-entry book keeping system.

trigger notice A colloquial term normally describing a notice by a landlord or tenant setting in motion a rent review procedure.

trust An arrangement that imposes on a trustee an obligation to perform specific duties in relation to the holding and controlling of property on behalf of and for the benefit of others known as beneficiaries.

trust corporation The Public Trustee or a corporation appointed to be a trustee or entitled under the Public Trustee Act 1906 to act as custodian.

trust deed, debenture A document securing debentures.

trust documents Documents held by trustees containing information in which the beneficiaries have an interest and which they are entitled to know.

trust for sale A trust under which the trustees must sell the trust property and hold the proceeds for the beneficiaries.

trust instrument The document that creates a settlement and appoints trustees.

trust property All property subject to the trust.

trust, breach of (See breach of trust.)

trust, charitable (See charitable trust.)

trust, completely constituted (See completely constituted trust.)

trust, executed (See executed trust.)

trust, executory (See executory trust.)

trust, express (See express trust.)

trust, power in the nature of Also known as a trust power or a power coupled with a duty. Created where the testator clearly intends the property to pass to objects whatever happens.

trust, public (See public trust.)

trust, purpose (See purpose trust.)

trust, sub- (See sub-trust.)

trust, termination of by beneficiary A beneficiary who is *sui juris* and absolutely can terminate the trust by calling for the property to be transferred to herself or to a third party regardless of the settlor's intentions or wishes. See Saunders v Vautier.

trust, void A trust which will not be enforced either because it is contrary to public policy or because it is illegal.

trust, voidable A trust which may be set aside because of some mistake.

trustee A person who holds and controls property in trust for someone else.

trustee *de son tort* A person who is not a trustee but acts as if she is.

trustee in bankruptcy In cases of bankruptcy, the bankrupt's property is vested in a trustee who must collect the bankrupt's assets, sell them and distribute the proceeds among the bankrupt's creditors.

trusts, unenforceable (See unenforceable trusts.)

turning The process of purchasing property and as soon as possible reselling it at an enhanced price. A process often carried out where the purchaser resells the property between exchange of contracts and completion.

turnkey deal The process whereby the vendor or lessor provides a building completely fitted out for immediate occupation. Hence the occupier simply turns the key and moves in.

turnover Total sales of an organisation for a stated period.

U

uberrima fides The utmost good faith, required in certain transactions, such as insurance contracts.

ultra vires Beyond their powers, especially of a limited company or statutory body.

umpire A third party appointed to adjudicate in a dispute where each party has appointed an arbitrator with whom they subsequently disagree.

unabsorbed costs Overhead costs of production which are not covered by revenue when output falls below a specified level.

unadopted road A road which the highway authority will not accept responsibility for as a public highway and which is therefore not maintained at public expense.

unappropriated profit Part of a profit of an organisation which is not paid out in dividends or allocated to any particular use or purpose.

uncertainty Where a will or deed is obscure and ambiguous so as to be incapable of being understood, it is said to be void for uncertainty.

under offer The situation where a property is subject to an offer which has been accepted in principle but the transaction still has to be completed.

under-capitalisation Situation in which an organisation does not have sufficient capital or reserves for the size of its operations.

underlease A lease granted by a lessee or tenant.

underwriter Term applied to a person or financial institution which guarantees to buy a proportion of any unsold shares when a new issue is offered to the public. Underwriters provide this guarantee for a commission.

underwriting The process lenders go through to evaluate the risks posed by a particular borrower and to set appropriate conditions for the loan.

undisclosed heir A person who claims the right to a piece of property after the death of an owner without a will.

undisclosed principal A principal whose identity has not been disclosed and is usually represented by an agent in negotiations to purchase the freehold or leasehold in property.

undisclosed spouse An unidentified marital partner who can claim the right to a piece of property.

unencumbered A title to land which is free of any encumbrances such as a mortgage.

unenforceable trusts Trusts which cannot be enforced usually because there is no human beneficiary (*cestui que trust*) to enforce them and they are not charitable.

unfair dismissal The dismissal of an employee that the employer cannot prove to be fair.

unfit dwelling A residential property which has been declared unfit by the local authority because of its failure to meet certain minimum standards of fitness for human habitation. See Housing Act 1985, as amended.

unfit for human habitation When the condition of a dwelling makes it unfit for habitation because of deficiencies in relation to repair, stability, freedom from damp, internal arrangement, natural lighting, ventilation, water supply, drainage and sanitary conveniences, facilities for the preparation of food and arrangements for disposal of waste water.

unit cost Costs incurred by an organisation expressed as a rate per unit of production or sales.

unit trust Organisation which exists for the collective purchase of shares and securities. Trust managers aim to minimise risk by spreading the investment over a large portfolio and so maximise profit for each individual unit holder.

unitary authority Local authority with responsibility for all local government functions including housing, social services and planning. All metropolitan councils are unitary and these are sometimes referred to as single tier authorities.

unlimited liability Liability to pay all the debts incurred by a business.

unliquidated damages Damages which cannot be calculated as a monetary loss, and which are assessed by the judge, such as damages for personal injury.

unlisted securities Securities, often equities, in companies that are not on the official stock exchange list.

unregistered land Land the title of which is not registered at HM Land Registry.

unsecured loan Any loan that is not backed by collateral.

uplift Additional rent payable when the terms of the lease give the tenant benefits not prevailing in the market.

uplift rent Rent payable when lease terms are considered more beneficial to the tenant than the usual commercial terms, e.g. a higher rent to reflect long rent review intervals.

upset price The reserve or lowest acceptable price. Usually applicable to auctions but in Scotland a term applied to transactions by private treaty.

upward/downward rent review A rent review covenant in a lease which requires the payment of rent on due dates, irrespective of whether this is equal to, greater or less than the rent payable immediately before the review.

upwards only rent review A rent review where the rent payable after a rent review is greater than that payable before it.

upwards/downward rent review subject to a base Essentially the same as upward/downward rent review, except that the rent cannot fall below a previously agreed level.

Urban Aid A subsidy paid originally through the Home Office, but then through the Department of the Environment to local authorities for work in urban areas. Housing associations may obtain funds through the local authority under this scheme for the provision of amenities and services to meet special social needs in the area in which they are operating.

Urban Regeneration Agency A government agency established in 1993 with extensive planning and purchasing powers.

urban renewal The redevelopment or rehabilitation of obsolete areas of a town, often including the provision of new infrastructure.

urban sprawl The unplanned expansion of development over a large area.

use and occupation action An action for damages for the use of property where there has been an owner and occupier relationship but no agreement for the payment of rent.

use classes order A statutory instrument defining which changes of land use need or do not need planning permission by virtue of the Town and Country Planning (Use Classes) Order 1987.

user One who uses, enjoys or has a right to property.

usury A reference to illegally excessive interest charged on any loan.

V

vacant possession Empty property which by law can be exclusively occupied or disposed of by the owner without any form of encumbrance.

vacant possession search A search undertaken immediately before completion of a property transaction to determine there has been compliance with a contractual term requiring that the property will be transferred.

vacation The periods between legal terms, when the superior courts do not sit. There are four vacations in the year.

valorem, ad (See *ad valorem*.)

valuation The act or process of estimating value; the amount of estimated value.

valuation certificate A certificate issued by a valuer stating the valuation of a property.

valuation list A statutory document prepared under the General Rate Act 1967 listing all the rating assessments of hereditaments in a valuation area.

valuation officer A public officer responsible for valuing property primarily for rating purposes.

value The price property will be expected to bring if sold at a particular time and in given market conditions.

value added Term to describe the value added to goods and services by each step in the chain of purchase manufacture and retail.

value added tax (VAT) Duty introduced under the Finance Act 1972. A tax payable by the ultimate consumer of goods or services. The tax payable is a percentage of the value of the goods or services sold.

value engineering Designing a product to eliminate any costs that do not contribute to the value of the product.

value in use The value of an item to the person using it which may be different from the sale price.

variable costs Costs which vary in proportion to the level of activity such as sales or production.

variable rate An interest rate that changes with fluctuations in such indexes as the Bank of England rate.

variable rent Rent payable under the terms of a lease which provides that it will change at specified dates by reference to a previously agreed formula or other means.

variance The difference between planned, budgeted or standard costs and actual costs. Variances are usually referred to as favourable or adverse, depending upon whether they increase or decrease profit.

variation (1) An amendment to an existing document. (2) The changes to the previously agreed design/specification made by the employer in a building contract.

variation order A written instruction from the architect authorising the contractor to alter or modify the building work. (See architect's instruction.)

velocity of circulation Calculation of the average number of times that a unit of money is used in a specified period which is approximately equal to the total amount of money spent divided by the total amount of money in circulation.

vendor Seller, particularly of land.

verge The external edge of a roof at the junction with a gable wall.

vertical integration Amalgamation of business organisations which are concerned with different stages in the production of goods and services.

vest The process of bestowing on another some legal right, or ownership of an interest in land or property.

vested in interest A term used to indicate a present or future right to enjoyment.

vested in possession A term used to indicate an interest which gives a right to present or current enjoyment as opposed to an interest vested in remainder which refers to an estate in reversion or remainder.

vested in remainder (See vested in possession.)

vested interest An existing right to a present or future interest in land. The interest vested may be 'in possession' or 'in interest'.

vestibule A small entrance hall or room.

vesting assent An instrument transferring ownership of land from the personal representatives to the beneficiaries under the rules of intestacy or a will.

vesting declaration (1) A declaration under the Trustee Act 1925 used on the appointment of new trustees, to vest the property in the new trustees. (2) When a compulsory purchase order has come into operation, the land may be acquired by a vesting declaration by the acquiring authority.

vesting deed A deed transferring the legal estate in the settled land to the beneficiaries in a settlement made during the settlor's life.

vesting order A court order by which property passes as effectively as it would under a conveyance, e.g. vesting property in trustees.

virement The movement of expenditure from one spread of expenditure to another, or from one financial year to another.

vocational training Training for a specific trade or occupation.

void Unoccupied, empty, unlet or unusable building or space in a building.

void allowance Sometimes called vacancy allowance, being a deduction made for the likely non-receipt of rent during a valuation exercise.

void contract An agreement which has no legal effect.

void relief Allowance against rates for the period when a property is unoccupied.

voluntary lien A lien a homeowner willingly gives to a lender.

voluntary transfer A colloquial term to describe the transfer of local authority housing stock to another landlord.

voting shares Shares in a company which entitle their owner to vote at the annual general meeting and any extraordinary meetings of the company.

W

wage freeze Government attempt to counter inflation by fixing wages at their existing level for a specified period.

waiting time The time during which employees are idle because they are waiting for work, material, repairs, etc.

waiver (1) A voluntary relinquishing of certain rights or claims. (2) The abandonment or failure to assert a legal right.

walking distress The list a bailiff prepares of goods to be distrained during the levying of distress but leaving them on the premises subject to an enforceable agreement that they must not be removed.

warranty (1) A term of a contract which is not a condition, especially a statement by the vendor as to the quality of goods. (2) An express or implied carrying out of the truth of a statement, whereby the warrantor becomes responsible in law in the event of the facts being otherwise than as stated, e.g. a vendor may warrant that property is fit for a particular purpose.

warranty deed Most valuable type of deed in which the grantor makes formal assurance of title.

wasting assets Property with a limited existence, usually refers to leasehold property.

wayleave A right of way over another person's land to lay cables, pipes or conduits, on over or under the land. Normally granted to a statutory undertaker.

well maintained payment Payments made by a local housing authority at its discretion for an unfit house which has nevertheless been well maintained. See Housing Act 1985, as amended.

wheelchair housing Dwellings designed for occupation by people who need to use a wheelchair.

white land Land which has not been designated for any particular use by a local planning authority. It is normally rural land.

whole blood The relationship between people descended from two nearest common ancestors.

will The legal declaration of a person's intention as to her wishes to be performed after her death. It must comply with the Wills Act 1837 to be valid.

will, conditional (See conditional will.)

will, international Will made in accordance with the Administration of Justice Act 1982 Schedule 2.

will, living (See living will.)

will, nuncupative (See nuncupative will.)

will, privileged (See privileged will.)

will, rectification (See rectification of will.)

will, republication of (See republication of will.)

will, revival of (See revival of will.)

will, revocation of (See revocation of will.)

willing seller/buyer/lessee/lessor An assumption normally associated with valuation purposes that the relevant party to the property transaction is willing to dispose or acquire his interest and that there is at least one genuine person in the market place for that interest.

wills, mutual (See mutual wills.)

winding up order Order given by a British court under the Insolvency Act 1986 requiring that a company be wound up.

without prejudice Correspondence in connection with a dispute, thus headed, is privileged and cannot be taken as implying any admission by the writer.

without reserve An auction or tender sale where there is no minimum price attached by the vendor.

words of purchase Words which describe the person who is to take an interest in land in his own right.

work in progress The value of work started but not completed. This will include partly finished goods, uncompleted contracts and unfinished products.

work study Technique used in the systematic examination of working methods designed to identify ways in which efficiency and effectiveness of work processes may be improved.

working capital The capital available for conducting the day-to-day operations of a business. This is usually defined as current assets less current liabilities.

working day For financial purposes a weekday excluding bank holidays.

writ of right Action to recover land in fee simple unjustly held from the rightful owner.

written down value Accounting term for the value of an asset less depreciation written off. This is not necessarily the same as its present selling price.

Y

year of assessment For capital and income tax purposes, a calendar year starting 6 April and ending 5 April.

year, executor's (See executor's year.)

yield The interest earned by an investor on his investment or bank on the money it has lent. Also called 'return' on the investment.

Z

zero based budget Budget in which responsible manager is required to prepare and justify the budgeted expenditure from a zero base.

zero bonds Bonds which carry no interest and are issued at a discount to investors who hope to make a capital profit on redemption.

zero rated goods and services Goods and services which are zero rated for the purposes of Value Added Tax.

zone A term commonly used in town planning being an area comprising a defined homogeneous land use.